PSYCHOLOGY OF AID

What can psychological insights contribute to aid?

International aid is immersed in the economic, political and cultural complexities which characterize relationships between the relatively well off and the relatively impoverished. These factors operate through the groups and the individuals engaged in aid work. If development projects are to succeed, it is clear that the human factors at work in the aid process must first be understood.

Psychology of Aid is the first book to focus exclusively on the applications of psychological thinking to international aid and the international aid process. Intended to help people think through development projects in a new and structured way, the book is split into four interlocking sections – discussing the dynamics of giving, delivering and receiving aid. Using case studies from both the less developed and more developed nations, the authors offer a wealth of psychological insights from the needs of the Northern donor to the tensions between Southern recipients of aid and foreign development agencies.

Illustrating that much international aid is psychologically constraining and, in the long run, counterproductive, the insights offered in *Psychology of Aid* seek to break the vicious circles formed by the current socio-dynamics of the aid process.

Stuart C. Carr is a Senior Lecturer in the School of Social Sciences at the Northern Territory University of Australia. His speciality is the social psychology of aid organizations, where he has researched and consulted extensively. Stuart is an Honorary Associate of the Asia-Pacific Institute of Human Resource and Development Studies, and Editor of the *South Pacific Journal of Psychology*.

Eilish McAuliffe is Director of the Masters in Healthcare Management program at Trinity College, Dublin. She has worked with various development organizations, including UNICEF, WHO, and the World Bank.

Mac MacLachlan is with the Department of Psychology, and is Director of the Centre for Health Behaviour, at Trinity College, Dublin. He specializes in health and organizational issues in international development, and has worked with many international and indigenous NGOs.

ROUTLEDGE STUDIES IN DEVELOPMENT AND SOCIETY

PSYCHOLOGY OF AID

*Stuart Carr, Eilish McAuliffe and
Malcolm MacLachlan*

Routledge
Taylor & Francis Group

LONDON AND NEW YORK

First published 1998
by Routledge
2 Park Square, Milton Park, Abingdon, Oxfordshire OX14 4RN

Simultaneously published in the USA and Canada
by Routledge
711 Third Avenue, New York, NY 10017

First issued in paperback 2014

Routledge is an imprint of the Taylor and Francis Group, an informa business

© 1998 Stuart Carr, Eilish McAuliffe and Malcolm MacLachlan

Typeset in Galliard by RefineCatch Limited, Bungay, Suffolk

British Library Cataloguing in Publication Data
A catalogue record for this book is available from the British Library

Library of Congress Cataloging in Publication Data
Carr, Stuart C.
Psychology of aid / Stuart C. Carr, Eilish
McAuliffe and Malcolm MacLachlan.
p.   cm.
Includes bibliographical references and index.
1. Economic assistance. 2. International cooperation.
3. Economic development projects. I. McAuliffe, Eilish.
II. MacLachlan, Malcolm. III. Title.
HC60.C2954   1998
338.9 – dc21   97–35288
CIP

ISBN 13: 978-1-138-86573-0 (pbk)
ISBN 13: 978-0-415-14207-6 (hbk)

We dedicate this book to our families — generations past, present, and future

CONTENTS

CONTENTS

TABLES AND FIGURES

Tables

Figures

PREFACE

Social analysts will always have to construct models, because it is necessary to set some issues aside to be able to comment on others. But they
should never lose sight of what they are doing and conflate the model
with reality, for in fact it is only a tool to get a handle on things. Moreover, no model, or hypothesis, or theory of social change is worth
much if it simply omits most aspects of *human behaviour* and makes no
reference at all to what people *think and feel*.

(Allen, 1992, p. 346, emphases added)

In several respects, this quotation summarizes what *Psychology of Aid* is about.
The book does not claim to have any kind of monopoly on international aid; nor
even for that matter on its *psychology* (for example, Marsella *et al.* 1996). But it
does set out to introduce the reader to a psychological way of thinking about aid
projects, while at the same time recognizing the need for complementarity
between disciplines (Rist, 1995). It is now clear that no understanding of aid can
be complete without at least some appreciation of the 'human factor' within it,
and surely only the most naïve of decision-makers would want to assume that the
psychological study of human behavior does not have something substantial to
offer in that regard.

Central to the experience of [participants] are psychological processes
such as those concerning motivation, attitudes, attributions, and perceptions . . . the new concept of development is one in which psychological
processes have a central place. By implication, there is a growing demand
for psychological knowledge in development projects.

(Moghaddam, 1990, p. 30)

International aid is surely part of this development effort, and the growing
demand that Moghaddam speaks of has motivated us to try and draw together
increasingly disparate aspects of psychological research and practice, often conducted by 'social scientists outside the domain of psychology' (1990, p. 29), into
one volume. In so doing, we are building on some excellent earlier foundations

(e.g., Gergen and Gergen, 1971, 1974), as well as anticipating that the wide range of topics will match the diverse requirements and interests of our readers. We have also referenced our points in detail, so that interested instructors and students can probe more deeply, if and when they desire.

A second factor that has motivated us to write the book, and one that is complementary to the first, is the growing disenchantment, including in some aid quarters, with the ethos of economic 'globalization.' As Eva Cox has argued in her Boyer Lectures, *A Truly Civil Society*, the notion of Economically Rational Man is threatening to destroy the very social fabric, or social *capital*, of our so-called 'global village' (Cox, 1995). This book, with its focus on the social dynamics of aid work, is broadly aligned with that recurring concern.

Like Cox's stance, our position on the civil society also runs counter to postmodernism, which 'seems to deny that we can identify injustice, or that we can act to prevent it' (1995, p. 5). Much of the 'torque' of the book is applied to the issue of social justice, and how international aid can work to add to, rather than detract from, the store of social capital. Such depletion has been happening, whether through backlashes deriving from failed projects (Salmen, 1995; see also, Gergen and Gergen, 1971) or through growing distrust between hosts and donors (e.g., Rist, 1995).

A major influence on us has been our common experience of living and working for several years in Malaŵi. Those years witnessed the failure of many well-intentioned aid projects, often, it seemed, because of the difficulties that human relationships and organizational behavior can breed. In this book, we describe much of our own research conducted throughout the 1990s in Malaŵi, integrating this with other research and theorizing. We do focus on the mistakes, but should not be taken as being unduly negative. Much aid 'works', and this is considered too; but we must surely learn more from our mistakes.

Drawing on research from and about one country is at once a strength and a potential weakness. As Cassen (1994) and Psacharopoulos (1995) have concluded, country case studies may be the only way of fully appreciating the interplay of factors that comprise the aid process. At the same time however, we also believe that our thinking and findings generalize to other contexts. Indeed, much of our own data (and those of others) resonate with strands in the wider development literature, drawn from many different and complementary disciplines. This book is therefore a modest attempt to raise consciousness about the value of a psychological perspective.

To many readers, explicitly adopting such a perspective will be a novel experience, and we have only used jargon where it serves to economically encapsulate a complex concept, which would otherwise require repeated wordy explanation. Striking such a balance is never easy though, and so, as an additional *aide mémoire*, we have appended a short glossary of terms. We hope that this will assist the reader in assimilating the more technical psychological terms, which we have tried to keep to a minimum. Above all, however, the book endeavors to stimulate, and that is the benchmark against which we would like it to be judged.

It should be pointed out that we have chosen to employ the term 'developing countries' rather than the more diminutive phrase 'Third World', or any similar euphemism. That is still very much a case of Hobson's choice, however, because all countries are developing. Further, our own view is that the geographical connotations, of developing *countries*, and even '*areas*' are becoming less and less appropriate as more and more variance appears *within* societies around the world, often between groups occupying the same physical space as each other. In some respects also, we regret the conventional terms, 'donor' and 'host', which also carry certain undesirable connotations (organ donation, disease carrier, etc.), although these at least convey the idea that aid is not always a welcome guest!

Last but certainly not least, we owe a great debt to, and thank accordingly, both students and staff at the Universities of Malawi, Dublin, and Newcastle (in Australia). Many of you have made invaluable contributions to the book. Deserving of special mention are Don Munro, for his consistently kind and worldly support, and Matt Hodgson, who has so vividly illustrated some of our central ideas. Mac MacLachlan is grateful for funding from Rotary International and Trinity College Dean's Fund, which partially supported his contribution to this book. Most of all however, we would like to thank our families, for their unstinting emotional support, patience, and encouragement.

1

AN AID CYCLE

The case

Zeffe is a Tanzanian national who was granted an aid scholarship from a Scandinavian donor country to study geology in Europe. During the course of his studies, he befriended a classmate from Sweden, named Bjorn. Throughout their studies and many group projects, they worked well together. Nonetheless, Bjorn tended to rely quite heavily on Zeffe, who was more senior and had much more actual field experience. Each friend eventually completed their studies and returned home. In accordance with his scholarship contract, Zeffe returned home and joined the State Mining Corporation in Tanzania as a geologist, on a salary of 36,000 Tanzanian shillings. This was during the late 1980s, when 250 shillings were equivalent to US$1. In US dollar terms therefore, Zeffe was earning approximately $144 per annum.

Soon after commencing work in Tanzania, Zeffe was told that an 'expert' was being recruited from overseas, someone who would head the Geological Mining and Surveys Department. This new Head of Department would, Zeffe was told, provide new and innovative ways of conducting business operations. The position was being funded and created by a Scandinavian aid agency. It would therefore be filled by an expatriate from the donor country.

Because the expatriate had to be recruited from the labor pool in a relatively wealthy economy, pay and conditions would need to be quite good, especially by Tanzanian standards. The salary would be US$24,000 per annum tax free, plus company car, housing, and several other fringe benefits. These included private schooling for dependent children, furlough (paid home leave), free electricity, and security guards.

Three months later, Zeffe was instructed to join the entourage sent to meet his new Head of Department at the airport. When the newcomer finally emerged from the Customs Hall, both he and Zeffe were visibly shaken. The new 'boss' was none other than Bjorn, Zeffe's long-time friend and (less competent) fellow student.

It wasn't long before Zeffe began to feel increasingly hurt and bitter about the rift that had been created between him and his former friend. His work suffered,

1

and he complained to his employers in the State Mining Corporation. They in turn protested to the aid organization, but to no avail. Zeffe soon felt compelled to resign his position, and he was promptly joined by several senior Tanzanian colleagues. Within three months of his arrival, Bjorn too had quit his job and returned home (Rugimbana 1996).

None of this powerful and moving case study is fiction. Moreover, as we shall see in later chapters, similar scenarios are still happening in other aid projects. In broad terms, this case illustrates what Sahara (1991) means when he says that the *relationship* between aid donor and host, such as expatriates and host counterparts, 'is now seen as a critical process, which can decide the success or failure of development intervention' (p. 6). Sahara's point is that the *human* factor is increasingly being recognized as a vital component in making aid work.

At the same time, we are witnessing a growing number of reviews about aid effectiveness (for example, Cassen, 1994; German and Randel, 1995; Picciotto and Rist, 1995; Wapenhans, 1993; World Bank, 1995a). This increasing interest in evaluation reflects growing *concerns* about the success of aid, at the project level and beyond. Indeed, over and above variations depending on sector and region, these reviews consistently show that aid projects often fail. As Cassen puts it, in what is possibly the most comprehensive review to-date, 'there is much that is already right with aid, but quite a lot to be improved' (1994, p. 14). Thus, even in Malaŵi, where aid is reported to have been relatively successful, 'the poor have not progressed and may even have got poorer' (1994, p. 40).

The argument developed in this book is that the failure of many aid projects may be connected to a comparative neglect of, or naïveté about, the so-called 'human factor'. Despite the convenient label, there remains a good deal of mystery in development studies about the contents of this particular behavioral box. As one aid consultant puts it, 'the development literature has been ominously silent when it comes to describing what actually transpires with a consulting team in the field. Like one's personal sex life, it is not a topic deemed worthy of public discourse. Yet it has engendered a rich, somewhat clandestine oral tradition' (Gow, 1991, p. 1).

The purpose of this book is to unravel some of that mysterious tradition, to begin to lift the lid from the human factor box. We will present new research and theory, focusing on *human behavior* in development settings. Much of the material is drawn from the discipline of psychology, which, until now, has not had a great impact on development practice.

Past

While psychological thinking has made some contributions to development, we believe that its potential practical value, in the area of aid projects, has not been fully appreciated or developed. By way of introduction, we must briefly paint the historical backdrop for the changes described by Sahara and Gow. In that backdrop there is a pattern. Whether we are dealing with Primary Health

Care (Stone, 1992), Human Geography (Craig and Porter, 1994), or Tropical Agriculture (Mwaniki, 1996), there is a cyclical pattern in the values that have often been promulgated as part of the 'culture' of aid.

Following the Depression of the 1930s, through the 1940s and 1950s, there was a somewhat patronizing 'silver platter' model of aid and associated development (Stone, 1992). During this 'do to' phase, as Mwaniki describes it, macro economists believed that the key to national development was imparting industrialization and modernization, from the fount of the donors' technical knowledge. *Human* factors were not particularly high on the aid agenda.

The 1960s and 1970s however saw themselves as the decades of 'cultural appropriateness'. During that stage, donors fell into two often conflicting camps. One party viewed traditional cultural practices as obstacles or constraints on aid-funded development. The other, led by early advocates of Primary Health Care (or PHC), construed culture as a 'resource'. For all such 'humanism', however, the PHC approach was still heavily patronizing and demeaning. To Mwaniki (1996), donor developers during the 1960s and 1970s had an unconsciously paternal 'do for' attitude toward their 'developees'.

The 1980s and early 1990s witnessed a return to macro-level policies. Monetarist and 'structural-adjustment programs' came to dominate theory and practice. During this period, economic rationalism became the primary beacon for development. This return to macro economics marks the first discernible cycle in the way that international aid has been conceived of and practised.

The 1990s and approaching millennium have in turn witnessed a growing sense of the power and significance of culture, an awareness that has undoubtedly been stimulated by the so-called 'postmodernist debate'. One poignant illustration of this growing awareness is the realization that aid itself often represents a form of cultural invasion (for example, Ingram, 1994; Mansell, 1995; Stone, 1992). Thus, in what amounts to a re-cognition, the 'human factor' is once more back in focus (for example, Adjibolosoo, 1995; Moghaddam, 1996; Ridker, 1994).

Given that psychological science is the study of human behavior, it would be surprising if thinking from that discipline was not generally included among the social sciences deemed relevant to international aid. Yet it may be argued that this is precisely what has happened. Although non-economic social sciences generally have been somewhat marginalized in development assistance (Klitgaard, 1995), this seems to be particularly so in the case of aid psychology. The Australian Agency for International Development (AusAID), for instance, does not employ *any* psychological advisors bar one token counsellor for visiting students (Jones, Reese, and Walker, 1994); while the British Overseas Development Administration (1995) excludes psychology from its list of 'fundable' disciplines (Carr, 1996b).

Such exclusion is ironic given the obvious scope for improvement in the success rate of aid projects. But more than this perhaps, we should ask *why* the study of behavior has become marginalized, to the point where it walks a tightrope between claims of irrelevancy on one side and neocolonial hegemony on the other. To answer that question, we must consider what went wrong the first time around, during that first humanistic stage back in the 1960s and 1970s.

During this period, at least one particular psychological notion seems to have captured the attention of social scientists; even to have created perhaps an indelibly negative memory. The notion in question is the so-called 'Need for Achievement', or 'N. Ach' (McClelland, 1961). Via a shared parentage in Western individualism (see Chapter 4, this volume), N. Ach is related to Maslow's (1954) 'self-actualization', an idea which remains ubiquitous in aid today (for example, Porter *et al.*, 1991), despite its arguably shaky foundations there (see, for example, Carr, MacLachlan, Kachedwa, and Kanyangale, 1997). Like Maslow, McClelland was a psychologist, who presented N. Ach as a psychological motive. The essence of this motive was seeking success 'in competition with a standard of excellence' (McClelland and Winter, 1971, p. 95). Thus, the high N. Ach person is driven by a need to compete, not only against others but also against her or himself.

Based on a wide variety of evidence, McClelland (1961) argued that this particular human factor was integral to national economic development. Moreover, it was supposedly learned, especially during cultural socialization practices. The implications were hard to miss. Owing to inferior cultural child-rearing practices, 'developing' countries had been deficient in the amount of N. Ach traditionally imparted to their young. By the same token, however, this could be remedied through (for example) aid-funded training projects. Thus, it seemed, McClelland had found both the 'constraint' on and the 'resource' for national economic development!

The difficulty we have with McClelland's formulation is that his *own* cultural values were taken as 'the solution', while those of his 'developing' counterparts were labeled 'the problem.' Here we see an example of how psychological notions about 'helping' at that time were inherently patronizing, without there being an explicit awareness of this. The whole notion of N. Ach itself was culturally inappropriate and insensitive. Many if not most cultures in the world do not value individual competition and self-promotion at the expense of others (Smith and Bond, 1993). Quite the contrary, in most nations *social* achievement is more the norm. Psychology's first foray into aid-related development efforts was therefore blinkered by its own behavioral assumptions; and the discipline, along with its way of thinking, have arguably been paying the price ever since.

Given what we now know about cultural definitions of achievement, it was inevitable that cracks would start to appear in the argument that more N. Ach would bring more economic development. McClelland focused much of his attention on training potential entrepreneurs to run small businesses. In such environments, as many economists today might assert, the motivation to compete is extremely important. McClelland's best-known study of the impact of training in N. Ach took place in India (McClelland and Winter, 1971). This aid project, funded by a succession of aid agencies, compared the impact of a course of training in N. Ach with a control group which had had no such training. An example of the content of such training would be an exercise in which the participants were asked to prepare an inscription they would like to see on their tombstone, thereby stimulating them to think about their life goals (McClelland, 1987).

Two years after the initial training intervention, McClelland and Winter found an increase in entrepreneurial activity. This was measured by a combination of business indicators, including number of new employees, capital investment, and profits. McClelland thereafter concluded that N. Ach could indeed be learned, and that it could indeed translate into sustainable economic development.

Yet the project had ignored sources of possible long-term resistance to these 'high achievers', stemming perhaps from cultural norms favoring community struggle over personal aggrandizement at the expense, literally, of others (Lawuyi, 1992). Thus it comes as no great surprise that on close inspection there are some worrying discrepancies in the data, anomalies that were reported by McClelland himself (1987).

Firstly, only those who had no superior managed to profit from the training. In Chapter 4 (this volume), we see how many societies place value on hierarchy and the preservation of status or 'face', which may result in a comparative tendency (see Chapter 9, this volume) to 'push down' on those who destabilize and threaten the stability of that system. Then again, it is also possible that simply being in a position to *implement* decisions could also have explained McClelland and Winter's findings. McClelland (1987) also points out, however, that those who respected *non*-N. Ach traditions, such as Rahukal (or taking a siesta), prospered *more* than their wholeheartedly N. Ach fellow trainees. Moreover, those individuals who did profit from the training were those who managed *not* to develop a desire to be recognized as a 'success'. In Chapter 9, we discuss the centripetal values emphasized in many cultures, and endorsed by many individuals within those cultures. These may motivate co-workers to 'pull down' individuals who are aggressively attempting to promote themselves above their peers.

In fact, the level of N. Ach in a culture has since performed very poorly as an indicator of longer-term macroeconomic development (Hofstede, 1984; Lewis, 1991). N. Ach has lost out to other 'human factors', such as Confucian Work Dynamism, a pattern of traditional Asian values that emphasizes thriftiness and long-term planning, and which was correlated highly (.70) with average Gross National Growth (GNG) in twenty-two different countries, including what was then rapidly developing East Asia (Chinese Culture Connection, 1987). Other research has linked (a reduced) proportion of Gross National Product (GNP) allocated to foreign aid with cultural values favoring personal acquisition and materialism (Hofstede, 1980). Such research does clearly indicate the value of psychology to development issues. It is therefore unfortunate that the baby (a psychological approach) seems to have been largely thrown out with the (N. Ach) bathwater.

Over a decade ago, Sinha and Holtzman (1984) assessed the impact that psychology had had in developing countries. Their review was not particularly favorable, suggesting that, far from aiding development, the discipline had been having very little impact (for example, Melikian, 1984), and could even have had a retarding influence on development (for example, Mehyrar, 1984). As we have ourselves suggested, this self-directed criticism would fit some of McClelland's aid-related work, on which so much seems to have turned. In what could have been an epitaph for the human factor in development projects, and helped along considerably by one of its own earlier advocates who took an extremely relativist position on psychological generalizations (Gergen, 1973), Sinha and Holtzman concluded that psychology had over-focused on person variables, at the expense of relevant contextual influences.

Constructs like N. Ach have also featured heavily in psychology's preoccupation with cross-cultural comparisons (Sloan, 1990; see also Gergen, 1973). Consistent with Sinha and Hotzman's review, these comparisons were sometimes insensitive to context, and thereby irrelevant to local issues: 'Comparing the citizens of New York and Nairobi, for example, is likely to be of no more use in developing Nairobians than it will be in developing New Yorkers'. An analogous point has been made about Abraham Maslow's 'hierarchy of needs' (Porter *et al.*, 1991). Our book breaks with those particular traditions, and the reader will find very little such psychology between its covers.

Finally, and again importantly for this volume, much of the earlier work was speculative and prescriptive rather than empirical and explorative (Sloan, 1990). This of course preserves the possibility that a human-factor approach could yet prove itself to be practical and beneficial in aid work. In the final analysis, this book is designed to document how that particular possibility may be starting to happen.

Present

Three particular areas of psychological practice are particularly relevant to development in the broad (Shouksmith, 1996), and thereby to development-related aid. Agreement on this is widespread, spanning Colombia (Ardila, 1996), Armenia (Jeshmaridian and Takooshian, 1994), Vietnam (Bazar, 1994), and Malaŵi (Carr and MacLachlan, 1996a). The areas concerned are social and organizational behavior, health and welfare, and education and lifespan development (Carr and Schumaker, 1996).

The following review of journal articles (i.e., excluding books and book chapters) from each area, plus accompanying summary tables, complete our brief survey of psychology's contribution to development. They represent the results of a search of the *Psychlit* database toward the end of the decade since Sinha and Holtman's last (1984) global review. No doubt, psychological and aid-related studies in other fields and countries are equally relevant, for example the more managerially oriented literature on leadership in developing countries (for example, J. B. P. Sinha [1990], which is incorporated in Chapter 4 of this volume).

Our own search terms were broad ('developing countr*' and 'Third World'), meaning that we did not restrict ourselves to studies that were exclusively and explicitly aid funded or aid focused. Many of the studies that emerged, however, *were* focused on aid, whether they were assessing the role of expatriate personality in projects funded by the Canadian International Development Agency (CIDA), surveying the therapeutic role of community support in schizophrenia on behalf of the World Health Organization (WHO), or evaluating early intervention projects for UNICEF. Moreover, insofar as all the studies were 'applied', and international aid generally seeks to have development impact on and through the human factor, the studies represented and tabulated below *are* relevant to the study of international aid.

Social and organizational psychology

A key social issue in recent years has been the role of WID (Women in Development) (for example, Kasente, 1996), and WID naturally remains high on the aid agenda too. Behavioral aspects to the issue of gender equality include findings that women employees may tend to develop more organizational commitment than their male counterparts (Alvi and Ahmed, 1987); that gaining paid employment often does not relieve the stress of domestic work (Gallin, 1989); and that self-employed women may be able to benefit from behavioral training in how to set difficult but locally realistic business goals (Punnett, 1986).

Commercialism may also be a prime motivator for prostitution (Penna-Firme *et al.*, 1991; McAuliffe, 1996), with some sought after women becoming the object of malevolent spells and Pull Down (see Chapter 9, this volume) to reduce their personal success (Kishindo, 1996). Such knowledge is potentially relevant to planning effective social marketing initiatives, aimed for instance at AIDS prevention in developing countries. Marketing, as we shall see, is inescapably relevant to aid organizations at all levels (McKee, 1992). These initiatives range from persuading the public to donate money, and promoting an image at home and abroad; to providing appropriate training and implementing acceptable health projects in host-community settings.

A somewhat fuller picture of the scope of studies in social and organizational behavior can be gleaned from Table 1.1. There are probably many more studies in locally produced journals and bulletins. This merely serves, however, to heighten the point of Table 1.1, which is to illustrate the need to draw together at least some of the wide-ranging psychological literature. That point is reinforced still further when we consider the much larger area of health and welfare.

Table 1.1 Social and organizational psychology articles 1985–94

	Regional focus				
Issues	Africa	Asia	Americas	Plural	Sum
Selection	0	0	0	2	2
HRD[a]	8	16	4	17	45
Occupational Health[b]	2	2	2	1	7
Marketing[c]	3	1	4	9	17
Modernizing[d]	1	1	0	10[§]	12
Intercultural relations[e]	3	0	1	6*	10
Interpersonal relations[f]	3	1	1	7	12
Totals	20	21	12	52	105

Notes:
§ This includes one study from eastern Europe.
* This includes one study from eastern Europe.
a In descending order: management development; achievement personality; technology transfer; organizational development; women in development; performance appraisal; ergonomics; vocational guidance; community participation.
b All concerned with women in development.
c Aid, market research, organizational seller behavior, market, alcohol promotion, dehumanizing effect of consumerism.
d Changes in: crime rates; locus of control; family life; child labor; marginalization; poverty due to arms expenditure; industrial status.
e Intergroup attitudes, including stereotyping influenced by Western media coverage of developing countries.
f Feminism; human rights; sexual jealousy; intimacy; interpersonal skills; attribution; belief in heredity.

Health and welfare psychology

Table 1.2 summarizes the health and welfare-related literature from 1985–94 inclusive. Social marketing clearly overlaps with health and welfare psychology, which is the most prominent category of psychological research conducted in developing countries. Examples of studies in this area would include: assessing risk behavior among aid workers themselves (Moore *et al.*, 1995); the impact of local perceptions on health projects (Nichter, 1995); predicting negative health and welfare correlates of industrialization (Cederblad and Rahim, 1986), combating malnutrition (Engle, 1996); rehabilitating refugees (Ager, 1996); and providing community support for the physically ill (Schopper *et al.*, 1996), as well as the mentally disturbed (Ustun and Sartorius, 1995). Community-support systems are also the focus of our own health-related chapters, in which we describe the psychological foundations for incremental (Chapter 5) and pluralistic (Chapter 10) social marketing of aid projects and goals.

Table 1.2 Educational and developmental psychology articles 1985–94

Issues	Regional Focus				
	Africa	Asia	Americas	Plural	Sum
Relevant role perceived for psychology[a]	1	1	0	12	14
Assessment[b]	8	6	3	11	28
Schizophrenia	2	2	2	22	28
Managing drug dependency[c]	1	1	1	6	9
Community health[d]	15	7	2	18	42
AIDS[e]	3	0	1	13	17
Industrialization[$f]	7	1	6	29	43
Nutrition and caregiver education[g*]	0	4	5	26	35
Abuse and neglect[h]	1	1	1	8	11
(Inter)national migration[i]	3	2	1	6	12
Refugees[j]	0	0	7	5	12
Intervening vs. disabilities[k]	1	3	0	8	12
Totals	42	28	29	164	263

Notes:
[$] Includes one study from eastern Europe.
[*] Includes one study from eastern Europe.
a Behavior modification, biopsychosocial model, psychotherapy, clinical psychology, health education.
b Tests and classification systems (often psychiatric).
c Trends and suggested preventative measures for abuse of alcohol and (much less often) use of tobacco.
d Buffering role for community and need for pluralistic forms of care.
e Often prescriptive, theoretical models.
f Physical (eg., diet and road accidents), mental (e.g., depression and suicide), behavioral (e.g., delinquency), and including both children and the elderly.
g Primarily nutrition, population control, and sanitation.
h Familial (child, parental, spouse), extrafamilial (child), physical, mental and sexual.
i Including streetchildren.
j From war, natural disaster and torture.
k Predominantly childhood, mental, and prescriptive.

Educational and developmental psychology

Educational and developmental psychology have possibly fared least well in developing countries (in psychology, 'developmental' refers to the lifespan-development process). In Table 1.3, the number of psychological studies in this group is by far the lowest, and serious doubts have been expressed about one of its mainstays (and a frequent yardstick for evaluating aid projects), educational testing (Zindi, 1996). Educational testing has been used elsewhere to identify children with special learning needs or handicaps, to identify especially gifted children, and to select pupils into further education. A number of indigenous

Table 1.3 Educational and developmental psychology articles 1985–94

Issues	Regional Focus				
	Africa	Asia	Americas	Plural	Sum
Testing	2	3	2	1	8
Literacy	1	2	0	2	5
Cognitive development[a]	2	3	1	5	11
Social development[b]	0	1	1	7	9
Socio-cognitive[c] and/or health	0	0	0	2	2
Development of school psychology	4	1	2	7	14
Participation of community[d]	2	0	1	2	5
Tertiary education[e]	1	0	1	3	5
Precocity	1	0	0	0	1
Totals	13	10	8	29	60

Notes:
a Models, early intervention, science and technology, and distance learning.
b Migrant adjustment; socialization, moral and emotional.
c Educational interventions generally; family influences on school achievement.
d Caregivers' expectations, cultural forms of childhood reaction to political repression, limits of Western models.
e Postgraduate selection policy and preparation for re-entry. Training for mental health care, teachers, and statistics.

psychologists and educationalists have recently called for more appropriate tests to be developed (Oakland, 1995), a call that has been echoed too in the area of literacy (Wagner, 1996). Such demand is now beginning to be addressed by test developers (for example, Baddeley *et al.*, 1995). A degree of success has already occurred with early intervention projects, which are reported to have produced sizable benefits in the socio-cognitive development of preschool children (Kâğitçibaşi, 1995), children in school (Sinha, 1990), and streetchildren (Taylor and Veale, 1996).

In summary, a brief content analysis of the literature during the mid-1980s to early mid-1990s indicates clearly that psychological research in developing countries is now widespread. Today, it has probably moved well beyond the simple three-way taxonomy presented in Tables 1.1 to 1.3. In an effort to grapple with that dynamism and complexity, the rest of the book is organized more in terms of a process or 'systems'-based approach toward aid projects. At its most basic level, systems analysis entails looking for recognizable patterns, in this case human-behavior patterns, or cycles, that recur in given situations over the course of time.

An '*aid* cycle'?

Donors at home

In itself, the idea of applying systems analysis to aid and development is far from new (for example, see Fountain, 1995). In discussing sytems 'archetypes', for example, Senge (1992) describes 'Shifting the Burden to the Intervenor.' In this system, a well-intentioned but outside intervention solves a problem so effectively that the recipient community never learns, or is never given the opportunity perhaps, to deal with the problem from within. Ultimately, that shortfall transpires because there has been insufficient initial focus on building a solution upon self-sustaining foundations within the host community itself. Instead, an attempt has been made to 'fix' the problem from the outside – say, by donors rather than by hosts.

Donors abroad

One of the most basic systems archetypes is known as 'Balancing Process with Delay' (Senge, 1992). We see this system operating when the shower water refuses to become hot and we impatiently turn the knob, not incrementally, but much farther, whereupon a scalding followed by a freezing fountain are each visited on our heads! The culprit is ourself, being less than fully aware of the delay between any corrective action and its effect on system output – in this case, water temperature. Another (all too familiar?) example is failing to take account of the delay between eating and feeling full, leading some of us to become overstuffed (Senge, 1992).

When the delay between action and outcome is relatively long, people may even 'give up' trying to influence outcomes, because they fail to see that any progress is being made. From this, Senge predicts that well-meaning, but vigorous, interventions will tend to produce instability in any sluggish system. Development, of course, cannot occur overnight, and the implications of this for aid-project evaluation, for instance, have been explored elsewhere (for example, Kumar, 1995). Our point however, is that local assessments of aid needs, made by expatriates in the field, may become what Cassen has described as 'sanguine' (1994, p. 105).

Hosts abroad

Social injustices are sometimes unintentionally created by sending expatriates overseas, on salaries comparatively high within a local context. These projects may then partly encourage hosts going to the West to opt for exile once their studies are completed, thereby creating spiraling demand for, and reliance on, expatriate aid. From the point of view of jobs and expatriates, this process may therefore develop into a system called 'Success to the Successful' in the sense that limited aid resources will be increasingly gobbled up by expatriate salaries rather than providing long-term investment in developing local human resources (Senge, 1992; for more 'macro'-aid-related examples of this particular archetype, see Dorward, 1996; or Fountain, 1995).

Hosts at home

Senge gives an example of food aid to developing countries, which may both lower death rates and increase population growth, thereby increasing demand for more food. Similarly perhaps, the erosion of some collectivistic extended families by wealth (see Hofstede and Bond, 1988) may shift the burden for care of the elderly to aid agencies (Senge, 1992, p. 61). In support of Senge's systems-theory-based proposition, Durganand Sinha (1991) has pointed out that this is now a genuine issue in developing countries.

In each of these scenarios, paying insufficient attention to 'the human factor' in the system means that technological assistance eventually precipitates a 'free fall into development'. Interventions by relief agencies may tend to shift the burden of responsibility to *themselves* (the intervenor), rather than to their intended target, of host institutional development. Behind this accidentally negative effect, according to Senge, may lie the problem of short-term, linear, and non-systemic thinking: 'If there is inadequate food, the solution must be more food' (1992, p. 63).

An alternative focus

Given that the application of systems theory to aid is far from new (see also Rondinelli, 1986), the reader may be asking, what is *unique*, if anything, about a psychological perspective?

One reply is that psychological analysis functions at the 'individual' level. This does *not* mean, as many have mistakenly assumed, that psychology itself is individual-*istic*. Individuals can, for instance, be motivated collectively (see Chapter 4, this volume). Nor does a psychological analysis rule out the influence of social factors, such as cultural norms, on individual behavior, thoughts, and feelings. But whereas sociologists, and possibly social anthropologists, will focus on relations between comparatively societal factors, such as class or religion, a psychological approach will tend to anchor observations to more micro-level factors, such as attitudes and specific actions that relate to those attitudes. For their part, economists might be more interested in larger-scale 'programs' than smaller-scale 'projects' (Kumar, 1995; Lawrence, 1989), whereas we, in this book, concentrate on the latter.

Let us also bear in mind that differentiation of focus does not mean that psychological thinking has any kind of 'monopoly' over any area of aid. On the contrary, the very spirit of this book is that different perspectives on the same aid phenomena may be complementary to one another. In fact, human-factor specialists have arguably had their greatest use in aid work when they have functioned as part of a multidisciplinary team (Wagner, 1996). Some of the best examples of this can be found in the context of South American community projects (Sánchez, 1996; Sinha, 1990; Sloan and Montero, 1990).

An interplay of systems theory and psychological analysis therefore forms the conceptual basis for this book. That is broadly consistent with the trend in management generally, toward thinking of organizations (and the clients they serve) as systems in which the 'human factor' is a major component (for example, Freed and Burack, 1996). Concurring with Cassen's conclusion that there has been a 'failure to learn from failure' (1994, p. 97), we describe an *aid cycle*, which is depicted in skeletal form in Figure 1.1. The essential elements of this systemic framework have already been introduced, and consist of donors at home (for example, NGOs [Non-Governmental Organizations] in the USA), donors abroad (for example, working as expatriates in India), hosts abroad (for example, aid scholars from developing countries studying in Australia), and hosts at home (for example, Malawian managers in Malaŵi).

This human-factor system portrays aid as functioning in a unidirectional fashion, in the sense indicated by the arrows. Such linearity should not be confused with how, or from which starting point, we think aid should ideally be working (see Chapter 12, this volume). Aid often begins (and ends) (Dorward, 1996; Ingram, 1994; Porter *et al.*, 1991) with the *donor* rather than the host at home. Among the donor motives listed by Porter and colleagues, for instance, are deflecting charges of domestic racism, gaining regional influence, and (as with our figurative examples in quadrants 2 and 4) 'flying the flag' for market purposes.

Our analyses have been designed to complement the excellent volume

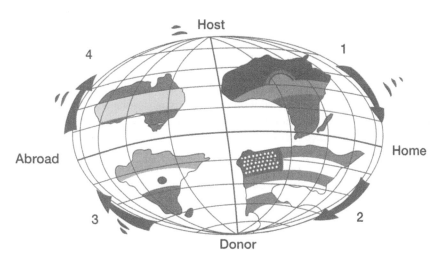

Figure 1.1 An aid cycle

Development in Practice, in several (psychological) respects taking up where Porter *et al.* (1991) left off. In the next chapter, for example, we extend their exploration of donor motives, considering donor publics and the task of soliciting aid funds from them. Non-governmental agencies (NGOs) frequently depend on media campaigns to fund their development projects. As such, they must effectively 'sell', or socially market, these projects to the lay public. Thus, they may wish to have some kind of understanding of lay public psychology with respect to social-marketing projects.

In practice, however, NGO project-aid media campaigns often rely largely on *their own* lay intuition about what tactics will have most psychological impact, in the faith, as Godwin (1994) observes, that the ends (donations) justify the means (crisis reporting). Until recently, most empirical research in this area had focused on demographic rather than 'psychographic' features of donors.

Studies have found that lack of direct experience of poverty tends to bias people toward blaming the victim, and thereby toward witholding aid. The sheer 'remoteness' of the so-called 'Third World', abetted perhaps by the 'victim' (rather than socio-political) focus of many media campaigns, might be exacerbating such a bias, for example, among those who believe strongly in a just (deserving) or an unjust (unforgiving) world. If so, some fund-raising efforts may actually be doing more harm than good, at least with regard to some major segments among the general donor public, both in the short term (regarding project donations) and in the long term (regarding intercultural relations). The undergirding idea here, that properties of an 'aid system' (like remoteness) may actually amplify social factors uncovered within Western contexts, is unusual.

The same point may apply to organizational decision-making, and the ways in which some aid agencies project themselves to their potential 'beneficiaries'. In Chapter 3, we consider the behavioral foundations of what Porter *et al.* have termed 'control orientation' in an aid organization. Allowing remote communities to set their own aid agenda would be an anxiety-provoking change for some aid agencies, and one that they therefore stubbornly resist even when projects clearly start to leave the proverbial rails.

One psychological factor here might be 'dissonance reduction', a process in which initial commitment can become its own worst enemy once a mistake is made (Dore, 1994). We have all 'bought' something or some idea that we could not really afford, or that didn't quite work as well as we anticipated, creating a discrepancy (dissonance) between what we know we should have done and what we actually did. Finding ourselves in such a quandary, we sometimes work very hard to maintain the belief that we acted correctly, often by convincing (deluding) ourselves that the product or idea is still a good one.

People in organizations are essentially doing the same thing when they refuse to admit that the candidate *they* chose for the plum job was the wrong one (Harrison and Bazerman, 1995). In some aid organizations, Gow (1991) observes that development models are seldom challenged by factual evidence of failure, and even provide a rationale for shifting the blame on to the victim.

14

Alternatives are excluded from the outset. Applying research and theory, we describe the well-trodden 'stepping-stones' to such 'groupthink' (Janis, 1982). Janis further devised ways of managing groupthink, and we consider some possible directions for enabling more meaningful participation.

Several authors now have drawn our attention to the length of decision-making chains in international aid, as well as its 'top-down' mode of functioning. Social and organizational research, too, have outlined the steps by which longer and more unilateral decision-making trees are likely to make messages, especially cross-cultural ones, more distorted (Bartlett, 1995; also, perhaps, McGuire, 1985) and less acceptable (Leavitt, 1951; Pruitt, 1995). The latter may be particularly so when the links in the chain form culturally distinct ingroups and outgroups (Turner, 1991), especially to traditionally communitarian societies (Smith and Bond, 1993). Little wonder that by the time aid finally arrives in the field it may have become an inappropriate, unhealthy, and imposing 'land-resettlement' project (Kloos, 1990; Porter et al., 1991), or a bitumen road on which, for hours, there are no cars and only one shoeless person walking (Grill, 1995).

From Figure 1.1, the theme that links together Chapters 2 and 3, and forms the first part of the book, is how donors at home project an image, either to the donor publics on whom they may depend for funding, or to the recipients on whom they depend for their very existence. Overlaid on this theme is the idea that properties inherent and particular to an international aid system, such as great distance and high uncertainty, may actually make much of the behavioral research and theory conducted within Western settings more, rather than less, relevant.

Once the donor has decided to aid a host, the next stage in the process is often sending someone overseas to work with the 'recipient' community or its 'implementing agency'. These expatriates are a key feature of many international aid projects, and in the second part of the book, from Figure 1.1, we consider these 'Donors Abroad'. As we shall see, expatriates can sometimes be as much a part of the problem as an element in its solution.

This problematic side to donors abroad often stems from basic differences in cultural values between donor and host, and in Chapter 4 we consider some of the key regards in which expatriates and local counterparts may differ. For example, while the deliverers of aid may (on the whole and in the long run) tend to value egalitarianism, individualism, materialism, and control, psychological studies have indicated dimensions in which some receivers of aid, in some developing countries, may traditionally stress the very opposite, namely hierarchy, group life, spiritualism, and tolerance of uncertainty (Hofstede and Bond, 1988).

Given such a discrepancy, there is a clear potential for donors' notions about 'participation' to translate into, or appear as, an 'imposition' from the cultural standpoint of the host (for example, Savery and Swain, 1985). We stress *potential* because there is often an overlap in values at the individual level. We shall discuss the eclectic labels that have been used to classify transactions between expatriates

15

and hosts. Underlying these terminologies may be a common dimension, reflecting the degree to which an expatriate is either task or relations oriented, with significant local ('emic', from 'phonemic') variations in what precise behaviors constitute each style (see Chapter 4, this volume). Managing a combined balance of each style, expressed in the culturally appropriate manner, is often the optimal leadership pattern. In India, for example, the fulcrum often consists of making good relations openly dependent on task performance.

A number of such preferred configurations, and thereby potentially the dynamic or systemic *interactions* with various expatriate preferences, are often quite well understood. In addition to providing structure for pre-departure training, such understanding could also be applied to select expatriates whose motives and values provide optimal fit, or overlap with, or tolerance of, the particular intended hosts.

Discrepancies between donor and host also figure in aid project expectancies, and in Chapter 5 we ask, what magnitude of change is likely to be optimal in the long run? Planned delivery, we argue, must consider the dangers inherent in large-scale changes to existing systems. Behavioral tendencies toward 'reactance' (anti-conformity) and social stability, for instance, may (eventually) counteract such changes. Furthermore, imported technology often creates human dependency on that technology, while locally sustainable resources correspondingly lose currency. The aid process fails to engage extant community systems.

Kurt Lewin's (1952) idea of a social force field in which, basically speaking, attempts at change may meet some resistance because stability (at least in some respects) is often preferable, reminds us to look instead for what is commonly termed a 'line of least resistance'. Less ambitious and low-technology projects, more in tune with the existing culture and circumstances, may make greater psychological sense. We introduce the psychophysics of incremental improvements and Just Noticeable Differences, and construct an approach to sustainable change based on learning to manage frustration. Such psychological thinking lends theoretical substance to Craig and Porter's (1994) warning that 'disengagement' is frequently the only way that hosts can otherwise preserve their social identity from 'drowning' in the (Trojan) 'gifts' of donors.

In Chapter 6, we consider these gifts in the context of the expatriate's motivation at work. Because of the very inequities that aid seeks to abolish, foreign workers are often paid from their home country. This, however, creates inequities with local counterparts, who are the intended replacements for the expatriates. Social and organizational psychology warn of serious dangers inherent in such inequities, such as derogation of the less advantaged counterpart and loss of intrinsic motivation. Those expatriates for whom this is most likely may actually be those who are most highly motivated to begin with, i.e., most sensitive to inequity. Thinking psychologically may thus inform selection procedures in rather non-obvious ways.

Other management alternatives include restructuring remuneration systems. The successful aid organization *Médicins sans Frontières*, for example, obliges expatriates to live and work under many of the same conditions as locals. It is also possible to consider the psychology of 'fit' between hosts and the increasing numbers of expatriates who originate from 'non-Western', and sometimes ethnically more similar, backgrounds. Surprisingly, again, we learn why and how similarity may partly hinder, rather than help, acceptance. Such resistance may be minimized, however, through pre-departure training. Maximizing the degree of fit between donor and host could thus be described as a linking theme for delivering aid via Donors Abroad.

The arrival of an expatriate often means the departure of a local person to gain qualifications in a donor country. In one sense, donors abroad in quadrant 3 of Figure 1.1 assume the duties of local persons and thereby release the latter for overseas study, in quadrant four. In another sense however, the expatriate temporarily 'displaces' an irreplaceable member of the local community (Dore, 1994). In Figure 1.1, the subsystems 'Donors Abroad' and 'Hosts Abroad' are inextricably linked. They form a (sub)system in themselves.

This link is further reinforced through the linking issue of fitting the person to the job (via selection) versus the job to the person (via analysis of training *needs*). When expatriates from 'non-Western' countries have been studied, the tendency has been to focus on their role as 'international students'. Given the relatively low numbers of 'non-Western' business-related expatriates in the past, this is partly understandable. What is much less acceptable, however, has been the implicit tendency to assume that these particular expatriates, being mere 'students', as well as mere visitors, should automatically assimilate to the West (Bochner, 1986).

In Chapter 7 we ask, who should adapt to whom? Perhaps the institution itself should learn from, and accommodate to, its guests? One consequence of assimilation, however, is an increased emphasis on improving selection procedures, and we describe several fundamental management techniques in that regard. Alternatively, we consider evidence that hasty commercialization of educational aid may prevent training from being tailored to local needs, with predictable long-term consequences.

The question of what happens after educational funding is withdrawn is posed in Chapter 8. In the past, answers to this question have often focused on so-called 'brain drain' (for example, Tan and Lipton, 1993), and we consider how the pushes and pulls on potential returnees can be considered as analogous, in some respects at least, to the Prisoner's Dilemma Game (see Dore, 1994). The concept of game-playing might be extended by envisaging it as one approach to understanding the *internal* exile that many returnees face. In the section entitled 'The case' (pp. 1–2), for instance, we saw how Zeffe's return was successfully negotiated in advance of his departure, but that this did not prevent his (and others') eventual alienation from an important job for which he had been professionally trained.

Other changes to the higher-education and training system might include building in more appropriate selection criteria, based on databases for the particular aid partnership in question. Alternatively, the training itself might be made more relevant to the host's home context, and its *particular* pattern of development needs. In both these tasks, the examination of archival data, differentiating the circumstances of returnees from those of exiles, may suggest psychosocial factors that 'make a difference' in that particular contingency. Overall therefore, the undergirding question for the section 'Hosts Abroad' in Figure 1.1 is 'assimilation or accommodation?'. The overarching theme is the application of business psychology to suit the increasing commercialism of educational systems.

One of the circumstances figuring prominently in any such question will be salary, namely the choice between equity in exile versus (as we saw in the section entitled 'The case') often gross inequity with expatriates back home. In the final section of the book, 'Hosts at Home' (see Figure 1.1), we turn directly to that most important of groups, the so-called 'local recipient' of aid.

In Chapter 9, we consider the psychological impact of salary inequities. Workers often react to comparatively unfair remuneration by reducing their own input, a reduction that is possibly amplified in proportion to the size of the differential! A sense of injustice might also be felt more keenly when cultural and community values stress comparative need over individual reward, factors that are again relatively common in host communities. Thus, multiple elements in the aid equation may sometimes combine to demotivate and disengage those who decide to return from their studies abroad.

As well as adding together, these elements can often push and pull in opposite directions. As globalization and its ethos of competition encroach further into aid and aided organizations, we may be witnessing a fundamental clash with traditional values that favor social over personal achievement, as well as stable hierarchy over a 'free-for-all'. Added to this, the growing emphasis on 'market forces' may be leaving many employees feeling job insecure, particularly so within economies that are defined as 'developing' to begin with. Promoting competition without regard for wider social context may thus be generating a variety of psychosocial conflicts at work, and we discuss the potential of traditional group and family metaphors to turn such conflict to collective advantage.

This theme of mobilizing traditional resources is continued in Chapter 10, which focuses on culture and community health. It is now becoming increasingly clear that a 'modern medical' monoculture is not automatically the most appropriate model for health organizations to pursue. We argue that aid organizations, instead of attempting to marginalize indigenous resources, such as traditional healers, may possibly increase their effectiveness by developing more inclusive services, and by regarding pluralism as a human 'resource asset'.

In the broader sense, this would simply be recognizing that the host has a *point of view*, a touchstone which provides the cornerstone for Chapter 11. In donors' eyes, some 'recipients' of aid may appear ungrateful and indolent, by asking for wages as well as aid itself. From a psychological perspective, however, this 'pay me to help myself!' reaction lays claim to a point of view. It represents protest against an imposing and demeaning form of 'assistance'. Like the donor's expect-ation of reciprocity, it is a demand for aid to become reciprocal, by *compensating* for a sense of lost cultural pride. Actively engaging with this negotiation process may mean that aid is less likely to sour, and ultimately is more likely to succeed. In Figure 1.1, understanding the dynamics of cultural resistance is the linking theme of 'Hosts at Home'.

The concept of cultural resistance now begins to suggest an overarching theme for the book as a whole. In the final chapter, we engage in reflection, seeking to extract an underlying principle from the diverse range of psychological evidence and theory reviewed. Resistance is usually a reaction to constraint, which in this case we identify as perceived social inequity, as each side locally defines it. In retrospect, the motivation to *restore* social equity can be seen as motivating the central events in our opening case study. That motive, on closer inspection, seems to run throughout the book. Thus we arrive at a theoretical capstone for the book, which is the advice to reduce constraint.

Mindful of the fate of N. Ach, this book makes no claim to have constructed some kind of psychological edifice for development. The failure of too many development projects can be laid at the foot of one academic dogma or another. Instead, we believe that the systemic and cyclical 'free fall' into poverty now being described by some observers is at least partly attributable to paying insuffi-cient respect to factors *human*.

Part 1

DONORS AT HOME

2

DONOR BIAS

A recent report from UNICEF (United Nations International Children's Emergency Fund) suggests that Western media images, including NGO marketing campaigns, may be seriously hampering efforts to persuade people to donate toward aid projects (Godwin, 1994). By accentuating the negative, as well as over-personalizing and under-emphasizing the socio-political causes of poverty (for example, debt repayments, famine), NGO fund-raising projects, like TV reports, might be helping to create a widespread and erroneous impression that the 'Third World' poor are a hapless, hopeless, and even blameworthy case. A similar warning has also been sounded within the West, regarding images of aid recipients in Ireland (Eayrs and Ellis, 1990). Such warnings potentially extend beyond the immediate, toward a more longer-term and perhaps universal erosion of social capital.

As a measure of its own concern over stereotyping the poor, and particularly in the international sense, UNICEF's Division of Information has been developing a program of learning exercises, entitled 'Education for Development'. This excellent package of exercises is designed to sensitize students (and therefore future donor publics) to the realities of life in developing countries (Fountain, 1995; see Hope and Timmel [1995] for a similar project in relation to extension workers). From the age of 10, Fountain warns, students in the West have often already formed negative stereotypes about persons from overseas.

Reducing such stereotyping is an area where psychology too may have a contribution to make (Gergen, 1994; Mehryar, 1984). We begin by offering a psychosocial analysis of UNICEF's concerns, finding support for them on both theoretical and empirical grounds. Synthesizing theory and research then allows us to derive some shorter-term, pragmatic suggestions for raising (1) public awareness about the real causes of poverty, and (2) donations toward aid project funds. Poverty eradication programs have tended to ignore the social construction of poverty by the poor themselves (Sinha, 1990b), whereas our recommended strategy centers on assessing and disseminating the perspective of actors 'in' poverty to observers 'of' it.

On a longer-term, and more theoretical note, the cross-cultural exchange of perspectives might also help to break an occasionally 'vicious circle', that goes something like this: blaming the victim instead of socio-economic forces → reducing donations to NGOs → helping to sustain an abiding image of a poor that cannot, or will not, help itself → blaming the victim once more. Meanwhile, the poor *themselves* may react to the way that they are being portrayed, developing an increasingly socio-economic view of the causes of both poverty and comparative wealth, coupled with a certain amount of indignation and withdrawal. An escalating communication gap therefore becomes distinctly likely, a process which could ultimately undermine public good will (social capital) on both sides.

Social cognition

One possible route for such prejudice to enter public consciousness is through the way that we habitually think about, and thereby explain the conduct and circumstances of others. The word 'cognition' derives from the Latin *cognitio*, for knowledge or understanding, and 'social' cognition is simply our everyday working understanding of people (including ourselves) and why we behave as we do. Probably the most studied topic in social cognition is attribution, or how people decide on the causes of others' (and their own) behavior (Heider, 1958).

According to Heider, attributions may be classified into two major categories, namely 'dispositional' (focused on personality traits) and 'situational' (focused on causes outside the particular person or people concerned). Blaming poverty on the poor themselves would count as a dispositional attribution, while blaming natural disasters, or international exploitation, would be examples of situational attributions for poverty, by referring to the situation people find themselves in. Blaming 'local corruption', on the other hand, could count as either situational (if the attributor separates the poor from their governments) or dispositional (if the attributor lumps everybody altogether into a single outgroup [Park and Rothbart, 1982]). Attributing poverty is a subjective and empirical issue.

A basic error

One of the most consistent findings from attribution research in the West is known there as the 'fundamental attribution error' (Ross, 1977). At the level of cultural generalization, people from Western societies tend to have a basic flaw in their lay theories about others' behavior. They tend to over-attribute it to dispositions, to the net detriment of equally valid, and often more valid, situational forces (Augoustinos and Walker, 1995). This bias raises the disquieting possibility that the donor public's attributions about poverty in developing countries might somehow be 'pre-primed' toward blaming the poor themselves.

The original demonstration of the fundamental attribution error indirectly supports that possibility. The participants (from the United States) read praiseworthy *or derogatory* articles about a developing country, Cuba, and were asked to describe the private attitude of the writer (Jones and Harris, 1967). Some readers were told beforehand that the writer had not had any real choice over the attitude adopted in the article. In this case, the writer's opinions were clearly explained by his constraining situation. Nonetheless, many readers doggedly inferred that the opinions expressed in the article were actually held by the author. Likewise perhaps, media audiences in the West often tend to see and believe that the 'Third World' is in perpetual crisis, and that its people are, in effect, a hopeless case (Dorward, 1996; Godwin, 1994). Perhaps this is partly because they take media commentators too much at their word?

An actor–observer difference

Much of the evidence for bias in our attributions about others comes from scientific comparisons with explanations that we give for our own behavior, i.e., when we are the 'actors' rather than mere observers of the behavior. Actors tend to make more use of situational attributions (Jones and Nisbett, 1972), for example, in explaining motives for charity work (Nisbett et al., 1973). Although both actors and observers can and do use each type of attribution (Augoustinos and Walker, 1995), observers (in Western societies, where most of the research has been done) are often less than mindful of situational factors (for example, Linton and Warg, 1993; Mitchell and Kalb, 1982).

A self-serving bias

The size of the the actor–observer difference varies according to whether we are attributing our own failures or successes in life (Nisbett and Ross, 1980). As actors, rather than observers, all else being equal, we can actually become over-focused on dispositions whenever our own success is involved. Could this also be so, we might ask, with *comparative* success, regarding one's own good fortune in comparison to, say, the poor? If so, Western audiences might in part be thinking rather too highly of their own abilities. In doing so, they could be implicitly downgrading people less fortunate, materially, than themselves. This might even be helped along significantly by the tendency, in some aid advertisements, to *personalize* the poor, by focusing on one person, by telling us his or her name, and generally by giving us as many personal details as possible. Such details could be encouraging viewers to blame the poor and take credit for themselves.

Lack of background expertise

A somewhat less 'self-serving' reason for the actor–observer difference has to do with background knowledge. The person who is in a bad mood is often the best placed to know why. Unlike casual observers, an actor has inside knowledge of the sequence of events, many situational perhaps, that led up to their angry outburst. An observer, on the other hand, is more likely to seize on the most 'self-evident' possibility (for example, 'she's just bad tempered!'), discounting the rest

(Kelley, 1972). Isn't this exactly the position of most Western TV viewers when suddenly observing fragmented images of the 'Third World?'

People watching

A further major element behind the tendency of observers to rely too heavily on dispositional attributions is the question of perspective. Physically speaking, actors from their vantage point 'see' the situation they are surrounded by, whereas observers are more likely to focus on the person in that situation. Theater lighting can be used to magnify this effect (McArthur and Post, 1977; Taylor and Fiske, 1975). With the same outcome perhaps, TV cameras often seem to regard the 'Third World' in precisely the same way. The 'eye' of the camera, usually operated by and broadcast to, a Western 'observer', is drawn to the human being, and hence – conceivably – toward ever-more dispositional attributions.

Figure and ground

This tendency to be drawn toward the human figure, rather than the situational background, has been nicely captured in Salvador Dali's picture, *The Three Ages*, which is reproduced in Figure 2.1. More than that, however, Dali's painting also illustrates what is known in perception as the principle of figure and ground. Our perceptual system tends to organize sensations so that more 'contoured' stimuli (for example, people rather than international exploitation) are given more substance and form (figure), while the rest are often relegated to the (back)ground of the perceptual field.

Figure 2.1 also shows us that what our perceptual system happens to class as 'figure' and 'ground' radically alters what we see and do *not* see. Even as far as human beings alone are concerned, we cannot see both figure and ground at once. One (often a person) is perceived, literally, at the expense of the other. Hence, after viewing numerous aid ads, that focus on people rather than on the wider context, we may begin to lose valuable vestiges of what Godwin terms the ' "balast" of back*ground* information' (1994, p. 47, emphasis added). When observing media events in the West, Westerners at least have access to *some* accurate background information in which to 'ground' their perceptions: 'We know that a riot in a city does not mean that the whole city, or the whole country, is rioting; we *know* that [it] is an unusual incident' (Godwin, 1994, p. 47). Much less so, perhaps, when regarding the portrayal of people in far-off and impoverished countries. There, 'figure' may be all that there is to rely on.

This rather alarming possibility has been brought home powerfully in Susan Fountain's work for UNICEF (1995). In *Education For Development*, one finds decontextualized pictures of 'pathetic' children, which are dramatically transformed and animated once the camera pans out to give, literally, the wider picture or background (in which the children are, for instance, actively attending to a school lesson). In our own experiences, too, there is a great disjunction

Figure 2.1 Figure and ground in Dali's *The Three Ages* © DACS 1998
Source: G. H. Fisher (1982). Ambiguous figure treatments in the art of Salvador Dali. *Perception and Psychophysics*, 2, 328–330. Reprinted by permission of the Psychonomic Society, Inc.

between the listlessness and sadness conveyed by narrow, fragmented media and NGO images of African children, compared with the vivacity and optimism that we observe in our everyday interactions with them.

Negativity bias

Godwin's example of rioting deliberately reflects the fact that reports and images from developing countries tend to be 'skewed toward the disaster end of the spectrum' (1994, p. 37). This raises the specter of what is termed the 'negativity bias' (see, Forgas, 1986). In management, for instance, just one negative feature on a job application, or in an interview, tends to be accorded a disproportionate amount of weight, easily overriding much positive and possibly more relevant information (Anderson, 1992; Herriot, 1991). In marketing, conventional wisdom has it that bad news travels farther than good news. In media coverage about developing countries, Dorward has described a tendency for 'crisis journalism' (1996, p. 4).

In social marketing of aid projects, Cassen remarks that, 'Public opinion seems more easily affected by one or two "horror stories" of failed projects' (1994, p. 224). It seems then that the negativity bias may be working *against* the needy. Ironically however, this particular bias may cut both ways. From interviews and surveys with aid officials from twenty-two different countries, US and Soviet volunteer workers tended to encounter great difficulties whenever their respective donor countries had been labeled at the time, in the media, as 'racist' or 'imperialist' (Gergen and Gergen, 1971, 1974).

Availability in memory

On both sides, such horror stories could possibly become part of what Senge (1992) terms 'reinforcing spirals' of negative attributions. Among donor publics, with repetition, negative images may become all that people remember; that is, they will form a negative stereotype in the memory. The 'availability heuristic' (Tversky and Kahneman, 1973) describes a tendency for people to over-rely, in constructing impressions, on what springs most readily to mind. Even politicians can be fooled (!) into over-estimating their chances of electoral success, because they remember only their own rallies and start to believe their own publicity (Gleitman, 1986). Victorious Roman generals knew better. On their triumphal parades, a slave would be whispering in their ear the Latin for, 'Remember you are only Human' (Cox, 1995). Among contemporary donor publics, moreover, it has been found that West Europeans grossly overestimate actual poverty-related statistics in developing countries (INRA, 1992, cited in Godwin, 1994, p. 46).

Donor bias?

There are undoubtedly wide variations among cultures, and among individuals living in those cultures, in how they make attributions and misattributions. But the sheer number of cognitive biases reviewed above leads us to one unavoidable conclusion: news and NGO campaign images from developing countries could, conceivably, be influencing Western donor publics to blame the poor themselves for poverty, rather than the situations in which they live. Aid advertisements themselves could be fostering 'donor *bias*'! (Carr, 1996b)

A possible retort

Even though aid advertisements may convey, perhaps unwittingly, the impression of dispositional causes of poverty (for example, lack of initiative, or intelligence), they might still manage to emphasize, more explicitly perhaps, the poor's lack of *controllability* over those circumstances (for example, through malnutrition or lack of schooling). 'If only you give, you can aid these poor people *past* those impediments!' (Meyer and Mulherin, 1980). Hence, from Figure 2.2., although aid ads may be missing the opportunity to steer attributions *directly* into situational attributions, they may, almost at the same time, be correcting or compensating for that 'wrong turn' by stressing the poor's (remediable) lack of blame for their predicament. Ultimately, that is, the viewer would be guided to the conclusion that the *situation is to blame.*

Social affect

This line of argument assumes that: (1) people are likely to be rather 'cerebral' when first encountering an aid appeal, calmly 'cogitating' about whether the poor could possibly control their own situation; and (2) that only *then* do they consider whether they should (rather more emotionally) blame the victim or not (Weiner, 1980). In Weiner's model, such emotion (or 'social affect') is then closely connected to (correlated with) giving help.

Yet, what if 'blaming the victim' is sometimes a more immediate affair, in effect 'short-circuiting' social cognition and moving directly to emotional, gut-reactions about whether to give or not? Affect, of course, is inherent (and indeed very primary) in many aid appeals, partly because aid is an inherently emotive topic, and partly because lay theory often has it that emotional appeals will enhance donations (Eayrs and Ellis, 1990; Thornton *et al.*, 1991). In social psychology, research by Forgas (1986) and others has established that there is often far more to attribution than just 'cold cognition'. Emotions, that is, often influence person perception directly and substantially.

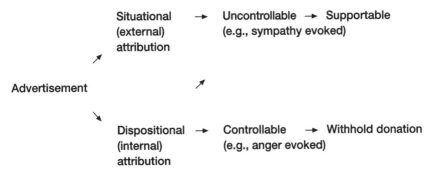

Figure 2.2 Possible pathways toward and away from donation behavior

The effects of mood

One particularly relevant finding is that negative moods like sadness (which many, if not all, aid adverts plainly evoke) tend to influence people to perceive others in a negative light (Forgas, 1986). This is particularly so when person perception is more complex and ambiguous, as (perhaps) in trying to make sense of images from far-off and unfamiliar lands, and when any kind of group difference is evoked (Tajfel, 1978). Examples of the latter would include the fairly obvious differences inevitably highlighted through aid advertisements (Eayrs and Ellis, 1990).

Defence mechanisms

Making negative judgments about, or blaming the victim, often serves an emotional protection function (Ryan, 1971). That is, victim-blaming functions as a convenient way of reducing feelings of guilt and helplessness in observers, for example, in relation to someone who is manifestly a lot less privileged than oneself, or has 'failed' in some comparative way. The relative deprivation of others, and our own sense of undeserved privilege, can sometimes come 'too close for comfort'. If, however, we can convince ourselves that these others have merely brought their misfortune on themselves (perhaps as a group) then our own comparative good fortune may become much less difficult (and even perhaps perversely gratifying) to bear.

If that is correct, we might expect most victim blaming to occur against those with the clearest misfortunes, including not only the poor in developing countries, but also cancer and AIDS sufferers closer (for Western viewers) to home. Indeed, in health promotion, for instance, bringing victims of AIDS closer to observers (of AIDS-prevention videos) has been linked to increased blaming of those victims for their (HIV) status (Berrenberg *et al.*, 1991). This is partly, perhaps, because viewers felt unable to help (Lerner and Simmons, 1966).

Privilege and proximity

In everyday life, many of us have encountered doctors who seem to have lost their 'bedside manner', to some degree because of repeated exposure to others' suffering. Some of us may also have found 'top' professors less than sympathetic toward their student 'minions'. Job 'burnout' is often associated with human service occupations, namely witnessing on a daily basis other people's suffering (Maslach, 1982). In organizational simulations, enjoying privileges over nearby others, including dispensing money to them, has been linked to developing a desire to maintain one's distance, as well as a certain derogatory contempt for them (Kipnis, 1972). That is especially so when one is more of a mere observer, and so feels relatively powerless to intervene directly (Lerner and Simmons, 1966).

Isn't that exactly what many people are made to feel, instantly, when witnessing highly distressing images of people suffering in close-up but out of reach in practical terms? Moreover, in an aid marketing campaign, a comparatively *passive*, 'countertop' solicitation (a small poster somewhat discreetly displayed at a supermarket checkout counter) was ultimately more effective than a scaled-up version that was also more actively 'in your face' (Thornton *et al.*, 1991). Presenting these (Western) consumers with photographs of children in need, as an adjunct to a charity appeal, actually *subtracted* from the effectiveness of that appeal for aid money (Isen and Noonberg, 1979).

Likewise, perhaps, Radley and Kennedy (1992) showed members of the donor public various photographs of needy recipients, and asked for opinions about their deservingness of aid. Perceived to be least deserving of all were the homeless 'on one's own doorstep'. Increasingly, one finds urban elites within developing countries and pockets of poverty (sometimes right alongside wealth) within 'developed' ones (Aroni, 1995). In Australia, Reser (1991) has noted acute actor–observer differences in domestic attributions for Aboriginal Australians' mental health.

What these studies collectively suggest to us is that potential donors may be most likely to blame the victim when that other's suffering is literally 'in their face'. Unfortunately, however, this is the very experience of combined privilege and proximity that many fund-raising campaigns deliberately set out to achieve.

Over-promising and under-delivering

Once we reconsider Lerner and Simmon's finding that a perceived inability to help exacerbates victim blaming, things begin to look even gloomier. First, the negative, as we have seen, is grossly accentuated through crisis journalism (Dorward, 1996). With repetition, this then gradually reinforces the suggestion that aid cannot solve the problem. Raising public emotions (Janis and Feshbach, 1953) and then showing how they can be alleviated (Leventhal, 1973), as in child-sponsorship ads ('There is terrible suffering, but all it takes to reduce that is a dollar a day') can be an effective way of changing behavior. In marketing parlance, it could be seen as analogous to under-promising and over-delivering (Forgas, 1986), as opposed to over-promising and under-delivering (Baldwin *et al.*, 1991). But the problem is that the ads do not *stop*. By sheer repetition, they run the risk of eventually creating the drained feeling that there is no hope anyway; in which case the promise of making a difference has been betrayed. It will have been transformed into *over*-promising and under-delivering.

This possibility has to be considered seriously when we learn that a major study of European donor publics revealed the following, disturbing sentiment: '*Giving money to Third World causes is like throwing it into a black hole*' (Godwin, 1994, p. 46).

An 'ultimate' error?

Connected to such negative sentiments is a bias that has been termed the 'ultimate attribution error' (Pettigrew, 1979). This is essentially the self-serving bias (denying responsibility for failure but grasping it for success) operating at a group level. It is especially likely to occur in social majorities of one kind or another (Augoustinos and Walker, 1995), such as (perhaps) members of donor publics taking the credit for their own comparative wealth, while implicitly or more explicitly (as above) blaming the poor for their own comparative lack of it. Tajfel (1978) has argued that the ultimate attribution error may be motivated primarily by the desire to *feel* superior to an outgroup, in which event it would be a process of social affect, working in tandem with the deleterious effects of privilege and proximity.

A second avenue for donor bias

To sum up, there are numerous reasons for concluding that attempts to raise aid-project funds, through emotive and essentially negative campaign images, may actually be complementing, rather than counteracting, a generalized donor bias. By repeatedly bringing the relative deprivation of the poor 'too close for comfort', fund-raising campaigns may sometimes be emotionally motivating many potential donors to distance themselves – psychologically – from the burden of guilt, toward blaming the victims of poverty for their own plight. In fact, the whole idea of simply bringing people into close contact with one another as a means of fostering intergroup cooperation (the so-called 'contact hypothesis') has never found much support in cross-cultural settings (Smith and Bond, 1993).

At the other extreme, however, as the psychological distance between the privileged self and the impoverished sufferer reduces (or fades in memory), the more 'cognitive' biases may come into play. These two extremes may have been captured nicely in Radley and Kennedy's (1992) study. They found that the two most undeserving categories, from among a variety of needy groups (for example, physically or mentally handicapped within the community) were perceived to be (1) the needy overseas and (2) the homeless on the doorstep. Perhaps, then, aid advertisements need to be wary of donor bias from two directions through allowing the poor to become too remote for sympathy; and by allowing them to come too close for comfort. Figure 2.3 indicates the difficult and delicate balance and coordination that NGOs are faced with striking in the media, if diverse peoples, from a wealth of 'developing' countries, are to avoid being stereotyped, in the public eye, into one negative social category.

Cultural contrasts

The 'fundamental' attribution error is not so fundamental or universal across cultures after all (for example, Hofstede, 1980; Ichheiser, 1943; Shweder and

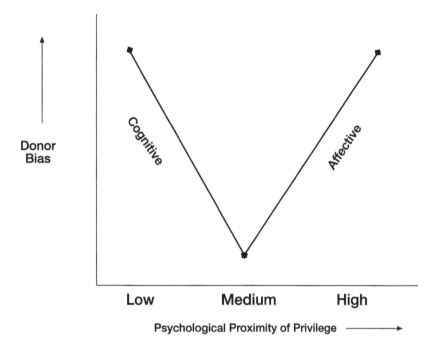

Figure 2.3 Possible effects of privilege and perceived proximity

Bourne, 1982; Smith and Bond, 1993), a point that probably applies to any attributional bias (see, Gergen, 1973; 1994; La Sierra, 1992; Monson and Snyder, 1977). At a cross-cultural level, Western-style socialization tends to emphasize individualism, whereas many (and perhaps even most) cultures in the world would tend to cultivate respect and deference toward the collective ingroup. These latter cultures, in their socialization practices, would probably tend to stress more the power of (social) situations (Markus and Kitayana, 1991). Indeed, while there may be few cross-cultural differences among children, adults from Western (versus 'non-Western') backgrounds often tend to prefer dispositional (versus situational) attributions (Korten, 1974; Miller, 1984; Miller *et al.*, 1990). Thus, the idea that children growing up in different cultures learn to explain the same events in different ways is now becoming more widely accepted (for example, Kleiner, 1996; Morris and Peng, 1994).

At a cross-cultural level, then, if people in developing countries are often more attuned to situational causes, might they not simply, as a group, blame their own poverty *too much* on their situation? This, after all, is what is suggested by the ultimate attribution error, which has a reasonably good track record across cultures (see Augoustinos and Walker, 1995; Smith and Bond, 1993).

To address the issue arising here, we can extend the analogy of actors and observers from within the West to encompass the *poor in developing countries as actors*.

An international actor–observer difference

As we have now seen, in a domestic sense, Western groups' actor–observer difference (and observer bias) is most acute when their own comparative success is involved. By contrast, 'non-Western' groups, being less individualistic, are often less self-serving. The societies in question tend to socialize their actors more toward modesty (Friend and Neale, 1972; Fry and Ghosh, 1980). Personal success is more likely to be attributed to luck or fate, and personal failure is often blamed on lack of personal ability or effort (Smith and Bond, 1993). To the extent that poverty is perceived as a form of failure, it would be *least* likely to be self-servingly blamed on situational factors, especially since disadvantaged minority groups, as we have seen, tend not to make ingroup-serving attributions.

Most of all, however, it is difficult for us to imagine a more 'outside' position to be in than a typical person, from a Western background, thinking about the causes of poverty in developing countries. More experienced, *inside*, observers have highlighted instead the often genuinely overwhelming power of situations occurring in developing countries (for example, Berry, 1979; House and Zimalirana, 1992; Lopez, 1987). They have thereby differentiated situations that are 'weak' from those that are inherently 'strong'. This distinction implies that outside observers, like TV audiences in the West, cannot help but underestimate the real strength of situations causing poverty in the developing countries (Ross and Nisbett, 1991).

Theoretical prediction

In any cross-cultural comparison of attributions for poverty in the developing world, we should expect to find *donor* bias (Carr, 1996b). If correct, this prediction would have some profound implications for the way that aid projects are socially marketed. We now present the first focused evidence that (1) what we have termed 'donor bias' may actually *exist* in a meaningful sense, and that (2) this donor bias may actually inhibit donation *behavior*.

Domestic poverty

There have been many studies of attributions of poverty. Most of these have been concerned with poverty within the so-called 'developed' countries, which is consistent with our earlier point that social divisions cut across geographical distinctions. The very first psychological study of poverty attributions was in fact conducted within the United States, by Feagin (1972). Approximately half the sample of 1,000 citizens believed that lack of thrift and effort, lack of ability and talent, and loose morals (all dispositional attributions) were very important reasons for domestic poverty. Situational causes such as low wages, prejudice, and bad luck were given much less shrift. Feagin was struck primarily by the degree to which these American respondents held the poor responsible for their plight, a tendency that unfortunately seems to have changed very little over time (Smith and Stone, 1989).

That, however, is exactly what we would have expected on the basis of the actor–observer difference. Most people in the United States at the time were probably not sufficiently poor to be counted as 'actors'. The actor–observer difference also predicts that the poorer minorities in the sample would have shown more signs of situational attributions, and this again is precisely what occurred. A small proportion of the sample was African American, and in the majority of cases (and in complete contrast to the wealthier majority) they made situational rather than dispositional attributions. Furthermore, among the sample as a whole, the tendency to make situational attributions generally increased along with reduced levels of schooling and lower levels of income, both of which are indicators of comparative poverty.

Feagin's findings have since proved to be remarkably robust. In addition to being replicated in the United States today, they have also emerged in Western Europe (Furnham, 1982a, 1982b; Townsend, 1979), in Australia (Feather, 1974; Reser, 1991), and in Canada (Lamarche and Tougas, 1979, p. 77). In a study involving nine countries of the European Economic Community (Commission of the European Communities, 1977), the two poorest countries (Italy and Ireland) were significantly more likely to attribute domestic poverty to social injustice.

In developing countries themselves, one study in India found that higher income was associated with a reduction in situational attributions and therefore a more dispositional emphasis overall (Singh and Vasudeva, 1977). Among the wider Indian population, however, situational attributions for poverty have been found to be used more frequently than dispositions (Pandey et al., 1982). These findings are again consistent with the notion of an actor–observer bias regarding attributions for poverty. Pandey et al.'s finding, in particular, contrasts with the general tendency toward dispositional attributions found within the wealthier West (Augoustinos and Walker, 1995).

This point of contrast clearly implies that there are differences between actors and observers in an *international* sense, namely between observers (and potential donors) living in the industrialized countries and actors living in the developing ones. According to our arguments, and the direct evidence so far, that difference would probably be due to bias on the part of the prototypical Western donor.

Poverty in the developing world

This donor-bias hypothesis has now been explored directly in several recent studies, involving attributions for Third World poverty as made by Australian observers on the one hand, and Malawian or Brazilian actors on the other. Each of these studies has made use of an instrument known as the Causes of Third World Poverty Questionnaire (CTWPQ).

The CTWPQ was first developed in Britain, by David Harper and colleagues (Harper *et al.*, 1990). The scale contains four independent, distinct subscales (Harper and Manasse, 1992). A dispositional factor is known as 'Blame the Poor' (for example, 'laziness', and 'keep having too many children'). Three largely 'situational' factors emerged. These are 'Nature' (for example, 'pests', 'disease'), 'Third World Governments' (for example, 'corruption', 'arms spending'), and 'International Exploitation' (for example, 'banking system', 'other countries') (see Cassen, 1994, p. 231, for support of the last two).

In a study conducted at weekend marketplaces in Australia and Malawi, 200 shoppers were interviewed using a locally adapted version of the CTWPQ (Carr, MacLachlan and Campbell, 1995). In this particular context, the scale items clustered into two statistically coherent categories, dispositional and situational. From Table 2.1, although on the situational items both groups tended to agree that situations were important, the Australians also made stronger dispositional attributions than their Malawian counterparts.

Donation behavior

In Feagin's original (1972) survey, increases in income and other indicators of comparative wealth were not only linked to dispositional attributions: they were also directly related to higher anti-welfare attitudes. That possibility has since been supported by subsequent research. For example, in the United States, Zucker and Weiner (1993) found that the more poverty was perceived as under the control of the poor themselves, the less it evoked sympathy and intention to help. In relation to aid for Kuwait and Iraq, attributions of responsibility were similarly crucial. Among nine Western European nations (Commission of the European Communities, 1977), those national groups attributing poverty to

Table 2.1 Shoppers' attributions for Third World poverty (mean scores per item) [a]

	Australian	Malawian	t-test
Dispositional	2.78	1.94	7.46 ***
Situational	3.14	3.18	n.s.

Note:
a Scale ranges from 1 (Strongly Disagree) to 5 (Strongly Agree). Therefore, anything above 3 denotes agreement with the cause.
*** Significant at the 0.001 level

Conclusion:
Since the Australian sample disagreed only slightly with a dispositional attribution on average, a sizable section of the Australian public are likely to be making the 'fundamental attribution error'. Australian donors, as individuals, were significantly more likely than non-donors to make *situational* attributions. Therefore, exposing donor bias may positively influence donation behavior.

Source: adapted from Carr *et al.* (1995).

sheer bad luck were also more willing to give their own money. In Australia, the most recent National Social Science Survey revealed the public opinion that international aid should be allocated to genuine victims of circumstance, not to those who can help themselves (Kelley, 1989). Disaster relief (caused by 'Nature' in the CTWPQ) was strongly supported, whereas 'aid for trade' was not.

Taken together, these findings all point toward one conclusion. Dispositional attributions of responsibility are linked to a reduced intention to help, while situational attributions are coupled to the opposite outcome. By implication, somehow *reducing* these dispositional attributions, and/or *increasing* situational ones (*up to* a point), may carry the potential to augment donation behavior. Indeed, in our marketplace study, Australians who actually donated to international aid organizations tended to make situational attributions, while their non-donating counterparts tended to ascribe poverty to the poor themselves (Carr *et al.*, 1995).

Some possible applications?

Given that a wide variety of donor publics in the West have been found to lean toward dispositional attributions, and that net situational attributions of responsibility may be critical for donation behavior (Weiner, 1991), how might NGO social marketers stand to benefit from empirical investigations such as the marketplace study? Attributional training (Martinko, 1994) has enjoyed some success in industrial contexts, and it may therefore be applicable to 'retraining' the general public to reconsider situational causes of poverty. This, of course, is also the basis for the UNICEF-sponsored work on 'Education for Development'.

We would like to suggest that the media could perhaps be used to *sensitize* the donor public to the realities of life in developing countries (Gergen, 1994; Mehryar, 1984). Presenting the hosts' (more situational) perspective *in their own words* might just be able to make a noticeable difference. In terms of UNICEF's objectives in their program 'Education for Development', members of the general donor public might 'become sensitive to stereotyping ... [and] learn to "read" the images presented to them in a wider context ... to see the individual or group behind the presented image as people with humanity equal to their own' (Godwin, 1994, p. 48).

There seems little doubt that the Western media can indeed influence people's attitudes toward developing countries (for example, Perry and McNelly, 1988). If only because of eternally competing considerations at home (Kelley, 1989), this would likely be a process of incremental rather than overnight improvement (for example, Beaman *et al.*, 1978). Nonetheless, Barbadians (for example) were reportedly more likely than their poorer Dominican neighbors (actors) to attribute Caribbean poverty to situational factors *after exposure to* an informative media that gave them a greater insight into the perspective of the Dominicans (Payne and Furnham, 1985; see also, Storms 1973, and Gergen 1994).

If the best we can hope for here is an *incremental* improvement, then why focus at all on 'psychological' factors? Conventionally, the social marketing of aid has relied mainly on demographic-segmentation variables, such as religion, social class, and gender (Griffin and Oheneba-Sakyi, 1993; Kelley, 1989). Some investigators, however, have applied regression techniques to compare the predictive power of such conventional factors against psychosocial variables, including attitudes toward overseas aid and attributions of responsibility for poverty. In the United States, for instance, political orientation was overshadowed by attribution of responsibility in regard to recommended aid allocations to Iraq and Kuwait (Skitka *et al.*, 1991). In Australia, welfare attitudes outpredicted demographic variables such as political affiliation and religion (Kelley, 1989). In fact, attitudes toward welfare there provided the best overall predictor of Australians' support (or lack of it) for foreign aid. Clearly, therefore, psychosocial markers could prove useful to social marketers of aid projects.

Belief in a socially equitable world: a pivotal motive?

From the vantage point of a personal belief in social justice, one is able to feel more comfortable about life's social inequities, such as poverty, by attributing them to the natural order (Lerner, 1980). In a just social world, you get what you deserve, and poverty in others becomes attributable to lack of personal effort, or some other failing on their part. By the same token, an innocent observer (perhaps oneself) is unlikely to fall victim to some nasty quirk of fate. In short, a belief in a just world often serves an ego-defensive function, helping the relatively privileged to make sense of, and reconcile themselves with, a 'big bad world'. A number of the studies reviewed above have obtained a link between the strength of this belief in a just world and more dispositional attributions for poverty (see, for example, Furnham and Gunter, 1984; Harper *et al.*, 1990). This belief may, therefore, partly motivate donor bias.

In a number of poorer countries (with perhaps less than their fair share of social equity), Furnham (1993) has found that people tend not to believe in a just world. The disjunction here between 'developing' and 'developed' countries not only concurs with the idea that collectivistic (and largely poorer) societies are less self-serving. It also raises the possibility that Western donors' belief in social justice may be one of the keys to donor bias. In fact, in times of need, just-world believers are often motivated not by the well-being of others, but rather by their own welfare (Zuckerman, 1975). Thus, in the Lerner and Simmons experiment described earlier, those participants who had stronger beliefs in a just world were more likely to *blame* the victim, while in our own student study in Malaŵi (Carr and MacLachlan, in press), as well as in Harper *et al.*'s (1990) survey of the UK public, belief in a just world was linked to donor bias.

At the other end of the continuum we found a possible link too, namely between belief in an *un*just world and (lack of) donation behavior. Those students who believed in an unjust world also tended, as a distinct group, to make

38

relatively dispositional attributions. Perhaps, in an unjust world, you also 'get what you deserve' because nobody else is left to take responsibility for you?

Furnham (1993) found that people living in more hierarchically structured societies tended toward greater belief in a just world *and* belief in an *un*just one. Clearly, then, gaps between rich and poor could be fostering both types of belief simultaneously – perhaps with the rich tending to cluster around one and the poor around the other?

From a somewhat more pragmatic point of view, and along with those who do not believe particularly in either type of world, there may conceivably be three new psychographic clusters, or market segments, for NGO fund-raisers to consider. Two of these, namely, belief in a just world and belief in an unjust world, may be linked, through victim blaming, to witholding aid-project donations. If so, aid adverts that focus on personally naming individuals, etc. and play down situational causes generally, could be encouraging donor bias and effectively 'switching off' *two out of three key marketing segments.*

Belief in a just world has also been linked to being conservative (Rubin and Peplau, 1975), which in turn has been linked with making more dispositional attributions for poverty, as well as with reduced intention to support international aid (for example, Gallup, 1972; Griffin and Oheneba-Sakyi, 1993; Kelley, 1989; Pandey *et al.*, 1982; Skitka *et al.*, 1991; Zucker and Weiner, 1993). What we may have here is a recurring configuration of beliefs, and social psychologists have a relatively long tradition of examining such patterns, especially as they relate to personality and victim–blaming (for example, Adorno *et al.*, 1950). In principle, therefore, such lines of enquiry could be rejuvenated somewhat, in order to attempt to unlock further the psychography of believing in social (in)equity.

A wider picture

Our donor-bias perspective is far from comprehensive. There are many studies on a great variety of 'human factors' in charitable behavior, as well as the social psychology of mass persuasion, and we would not wish to understate their potential relevance or value. To take just one example, much research has been directed at the added value, compared with simply asking people for money, of having a 'foot-in-the-door' (Freedman and Fraser, 1966) and even a 'door-in-the-face' (Cialdini *et al.*, 1975). In the former technique, an innocuous request (for a small donation) is followed by a request for a larger one. In the latter, an unrealistically large request is followed by a much more modest one.

These techniques work especially well with requests for help (Dillard *et al.*, 1984). Each technique has also demonstrated an improvement in donations compared with simple requests (Cantril and Seibold, 1986), whether for charitable donations (Wang *et al.*, 1989), or for people to volunteer (Cantril, 1991). Thus, these techniques have already indicated their potential value, at a 'quick-and-dirty', pragmatic level, in augmenting donations toward aid projects. Each technique has a fairly clear theoretical foundation (see De Jong, 1979; Goldman, 1986; Grace *et al.*, 1988).

A potentially more robust (though still Western-origin) model, however, is the Theory of Planned Behavior (Ajzen, 1985). Earlier versions of this model have been successfully applied to predict altruistic, for example, blood-donation behavior (for example, Bagozzi, 1981; Burnkrant and Page, 1982; Charng *et al.*, 1988; Zuckerman and Reis, 1978). In the current model, and probably in others (for example, Triandis, 1980), whether one will donate is best predicted by intention to donate, which in turn is predicted by a combination of attitude toward the particular charitable behavior (strength of feeling 'for' or 'against'), subjective norms (the various *social* pressures that one feels to act charitably or otherwise), and practical issues of control (for example, 'can I *afford* to give?' 'Will the money get *through*?'). The latter is a recent addition to the model (and remains to be tested in relation to aiding others), but each of the other two has been found to contribute toward students' intention to bequeath money (Konkoly and Perloff, 1990). Thus, attributions for poverty may provide just one element in the attitude component, which in turn might also consist largely of deservingness (Feather, 1994), or (more likely perhaps) *relative* deservingness (Hull, 1988). The practical value of a theory like planned behavior is therefore that it begins to suggest to social marketers where they might target their campaign messages, in order to obtain maximum impact.

Traditional factors

As well as needing to assess the value of models like the Theory of Planned Behavior, it is also important to realize that they can never be 'comprehensive' (Rugimbana *et al.*, 1996). That is, they never manage to account for all the variation in people's behavior, or even in their intentions. They are, after all, only *models*, and *should* be simpler than reality. However, one extra factor contributing to this shortfall, in Western-style models, could be their emphasis on 'cold' cognition. 'Planned behavior' even used to be termed 'reasoned action'. Perhaps, then, we might begin to reconsider more social-affect factors, such as one's sense of moral and social duty (for example, Triandis, 1980)?

Such possibilities were certainly envisaged by Azjen himself (1991, p. 199), and in support of that prospect, where the model has been tested cross-culturally, and on topics that are not relatively private and personal to begin with (Godin *et al.*, 1996; Matsuda, 1985; Wilson *et al.*, 1992), the role of subjective norms has actually been shown, as one might expect, to be relatively high in more collectivistic settings. The theory may therefore be of some value cross-culturally, being flexible enough to accommodate to, capture, and *reflect* the impact of traditional values communicated to Western audiences.

What traditional factors?

Given that most poverty-attribution research and theory has been focused on domestic inequities in the West, the existing measurement scales do not afford full and proper representation of the non-material type of explanations that have often been used, and continue to be used, in non-Western societies (Augoustinos and Walker, 1995; Evans-Pritchard, 1976, respectively). Because the CTWPQ was originally designed for application in exclusively Western settings, it was inevitably developed solely from the perspective of Westerners (Britons originally) rather than that of the poor in developing countries.

In cross-cultural work, that in itself is liable to foster bias. In Indonesia for example, sheer luck is seen as (and may partly *be*) more important than in the United States (Marin, 1985). In Malaŵi, and on the CTWPQ, the sole clearly non-material item, 'because of fate', has never correlated with any of the major themes derived through factor analysis. The statistical independence of this item is intriguing. It indicates that the scale could be developed further, to provide better representation of non-material beliefs. That in turn means that there is potential to enlighten Western donor publics, namely by giving them some insight into the 'fatalistic' perspective of the poor themselves. Attributions concerning fate, chance, or spirits might also be considered 'situational' (or at least beyond 'human' control), and to that extent possibly helpful with regard to augmenting donations.

A pathway in that direction may already have been embarked upon, both in Brazil (Tamayo, 1994) and in Indonesia (Sarajar and Soemitro, 1997). Tamayo, for example, has found that the primary factor explaining Brazilians' attributions for slum poverty was belief in 'God-destiny'. If research like this continues to progress, we believe that presenting such perspectives, in greater detail, could enhance the general impact, on the donor public, of having poverty described from the perspective of the poor themselves. A contextually widened theory of planned behavior, with its emphasis on assessing the weight of each component in donation behavior, could then serve to quantify the impact of donors experiencing a traditional perspective.

A systems-theory integration

Persuading people to 'throw money at' the profound social problem of poverty is not the only concern. What about the longer-term and wider social dynamics of comparative poverty and wealth?

It has been argued that socio-cultural (Smith and Bond, 1993) and socio-economic factors (Hofstede and Bond, 1988) may predispose Western donor publics toward individualism. According to our arguments, that means a relative propensity to blame the poor in developing countries for their own poverty, a leaning that may become 'activated' by NGO fund-raising and news-media images of developing countries. Simultaneously, the socio-cultural and socio-economic context of the poor, we have suggested, may incline *them* to see their poverty as beyond their own control. That judgment, we noted, may often be veridical, or at least *relatively* accurate. Former Zairean President Mobutu's personal fortune, for instance, was frequently estimated at billions of dollars, rivalling the entire national debt.

Taking this dynamic further, in Britain, Adrian Furnham (1983) found that Labour voters were more likely than Conservatives to attribute *success and wealth* to situational factors, like luck and chance. As Furnham remarked, this finding mirrors the pattern found with attributions for poverty. More importantly how-ever, these data remind us that there may be more to attributions about poverty

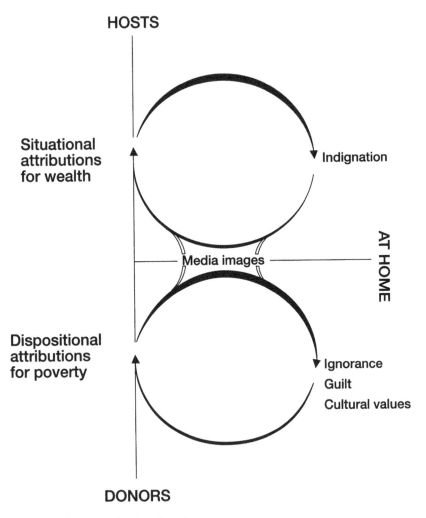

Figure 2.4 A communication chasm?

than simply attributions about *poverty*. In other words, *the poor have a point of view about the (relatively) rich*. This point of view may in turn interact with the way the poor's own poverty is portrayed in the world's (Western-dominated) media. Many villages in the developing world today have regular access to these media and their images. Systems theory suggests that their point of view will eventually somehow input into those media portrayals themselves, thereby 'interacting', via the mass media, with the views of Western audiences.

From Figure 2.4, as Western media images are increasingly seen by both donors and hosts, they may be contributing toward an escalating communication gap and ultimately, perhaps, a wealth gap. In our theoretical system, one side may occasionally react to media images by blaming the victim (and withholding donations), while the other group/side is left feeling somewhat demeaned and, perhaps, somewhat indignant. As the portrayals and associated reactions wear on, they may actually begin to feed back *into* those very portrayals and reactions themselves, thereby augmenting their effects. Donor publics' finite store of good will is steadily undermined by the continuous stream of negative images, so that donor bias will increase and donations will decrease. On the host side, when people are repeatedly stereotyped, their behavior has a habit of increasingly appearing to fit the stereotype more and more – the self-fulfilling prophecy or 'Pygmalion effect' (Merton, 1968; Rosenhan, 1973; Rosenthal, 1968).

Indirectly, and in the long term, the escalating misperceptions in Figure 2.4 may even begin to contribute toward a more macroeconomic cycle, namely 'the rich get richer and the poor get poorer' (Senge, 1992). On the donor side, one of the corollaries of economic and material 'development' (as we have seen) is increased individualism, and hence – maybe – increasing donor bias. Thus, the process of escalation that we have described in Figure 2.4 may start to have an incremental input into the wider economic interplay between developed and developing countries (Dorward, 1996).

Whatever the case, one of the benefits of adopting a systems-theory perspective is that it reminds us to look beyond personal events. From this inherently *less dispositional* perspective, the media bridge depicted in Figure 2.4 could surely function much more productively? According to Senge, one of the best ways to break such circles is to expose the biases of each side to one another. That, in essence, is what we have been proposing throughout this chapter.

3

THE AID CHAIN

A factor contributing to the problem in the case study in Chapter 1 is that aid decisions were made remotely from the recipient community, without any real consultation at the local level. The result was an inappropriate and impositional aid project, which soon provoked simmering resistance, and eventually outright failure. Almost by definition, such 'remote-control' scenarios are partly embedded in the very concept of 'international' assistance, and may also be partly encouraged by relative tendencies in Western cultures (Hofstede and Bond, 1988), including aid organizations like the World Bank (Vittitow, 1983), toward avoiding uncertainty and maintaining control. Referring specifically to aid agencies, Vittitow, for example, remarks that 'traditional development strategy emphasises certainty, "hard data", and controlled variables' (p. 309). Vittitow's analysis focused on a large multilateral aid organization, but Porter and colleagues (Porter *et al.*, 1991) have extended the concept of 'control orientation' to include NGOs. In management terms, Fowler (1996) has written an excellent description of top down 'linear' tendencies in what he terms 'the [NGO] aid chain'. This chapter builds on those previous works, by offering a psychosocial analysis of the dynamics of making decisions along the aid chain.

The aid chain

Craig and Porter (1994) have pointed out that a typical NGO project will involve a funding body, the donor NGO and its consultancy firm, a partner intermediary NGO in the host country, and several other levels of local organizations. This 'chain' of organizational decision-making groups, as they term it, is schematically portrayed in Figure 3.1.

According to Craig and Porter (1994), each of the links in a chain will recast the aid project 'in terms of their own concerns and priorities, and practical capacity . . . before its objectives can become visible in and to the target population' (pp. 19–20). Therefore, at each link in the chain, there exists an agenda which may or may not be consistent with agendas in the links both above and below it. These inconsistent and often conflicting agendas that exist at different points in the aid chain create difficulties in the determination of

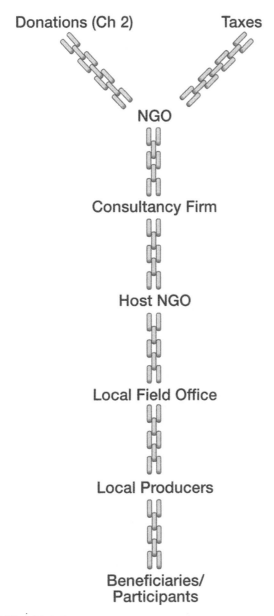

Figure 3.1 The aid chain
Source: adapted from Craig and Porter (1994) and Fowler (1996).

expected outcomes for any particular program or project. In the absence of well-defined and meaningful outcomes, the effectiveness of programs cannot be accurately evaluated.

Establishing an agenda

Inconsistent agendas

Inconsistent agendas arise from differences in understanding of the problem, or different approaches to tackling poverty alleviation. Nowhere is this more obvious than in the controversial International Monetary Fund (IMF) lending policies, with their stringent conditions. The IMF, although not an aid agency, has become increasingly involved in the aid picture, particularly in low-income and in less developed countries. The IMF was created for the purpose of providing loans to its member states facing balance-of-payments deficits. It has been inclined to impose severe economic austerity programs upon African countries with serious payment deficits. These programs include stringent curtailment of public expenditures, restrictive fiscal and monetary policies, devaluation of the nation's currency, removal of state subsidies and removal of the restriction on the free flow of trade and foreign investment (Anunobi, 1992).

The term 'structural adjustment' is used to describe 'the kinds of policies now recommended to developing countries that are heavily indebted, or, for other reasons, have balance of payments deficits with the aim of enabling them to pay their debts, to become credit worthy again, and to lay the foundations for subsequent sustainable growth' (Singer, 1991, p. 41). Economic structural-adjustment programmes (SAPs) often call for changes that entail at least short-term sacrifices by those who are affected by them. To date, SAPs have concentrated on reductions in public expenditure more in the social services than in any other area (Rahman, 1991). This has resulted in major sacrifices being requested of those that are poor both economically and in terms of their health. Understandably, the developing countries in receipt of these loans do not always agree that such sacrifices are justified or necessary and some disagree fundamentally with the policy of cutting public expenditure in the social-services areas.

However, World Bank structural-adjustment loans (SALs) have become virtually conditional on the recipient country's agreement with the IMF on a fund program (Cassen, 1994). Conditionality was perceived to encourage policies that would make it more likely for a member country to cope with its balance of payments problem and to repay the fund within four to five years (Anunobi, 1992).

There have been signs, particularly in relation to some East African countries, of differences between the World Bank and the IMF, reflecting to some extent their differences of objective: IMF being concerned with short-term balance-of-payments adjustment; and the World Bank with longer-term developmental considerations (Cassen, 1994). Many African countries have frequently relied upon the IMF, but they feel that IMF policies usually ignore economic and political realities and place unwarranted burdens on their economies. Thus, they rely upon the IMF only as a last resort. The deflationary policies imposed by the IMF produced political upheavals in less developed countries, including African states, referred to as 'IMF riot'. Domestic chaos confronted regimes in Turkey, Peru, Portugal, Egypt, and Chile during the 1970s in the wake of these governments' acceptance and implementation of austerity packages required for access to IMF loans (Payer, 1975). Countries such as Tanzania, Zambia, and Jamaica have found the conditions attached to IMF lending so objectionable, economically, politically, and ideologically, that they have broken off negotiations for desperately needed foreign aid. While most less developed African nations have not gone this far, they all find the intrusion of the IMF upon their domestic and foreign policies both unhelpful and unacceptable. The problems created by these differences between the IMF and the World Bank, and between both of these and aid-recipient countries, illustrate the negative effects of tension at various points along the aid chain. These tensions are exacerbated when one link in the chain imposes its agenda on another, as in the IMF's imposing conditions on countries in receipt of IMF loans. The recipient country is thus compelled to comply with the IMF agenda, even when this conflicts with the country's own internal agenda or policies.

Donors are often accused of being driven by internal political or bureaucratic objectives and constraints, rather than by the needs of the recipient countries. In some instances, domestic political pressures have forced the donor agencies to oblige the recipient country to purchase all expertise and equipment in the donor country. This procurement-tying is particularly prevalent in bilateral aid arrangements. It frequently results in non-competitive pricing of donor goods and services, thus squandering resources and undermining effectiveness. It almost invariably lessens the chances of sustainability, as the recipients remain dependent on the donor country for equipment parts and servicing, even after the donor agency has terminated the project. This clearly illustrates that there is often a substantial difference between the imposed donor agenda and the agenda of the recipient country, again leading to tensions in the aid chain.

Fundamental differences also exist between NGOs and their funding agencies. Edwards and Hulme (1994) explain how a central component of what has been referred to as the *World Bank's new policy agenda* is an ideological preference for markets over state controls in the allocation of scarce resources (which could be loosely interpreted as a preference for private- over public-sector enterprise). This ideology runs counter to practice within international NGOs and official aid agencies, i.e. the provision of subsidized credit to particularly vulnerable social groups in an attempt to accelerate economic growth and reduce poverty. Copestake (1996) focuses on the issue of subsidies to demonstrate the differences of opinion that exist not only between the World Bank and NGOs but also within the World Bank itself. Pro- and anti-subsidy positions lead to differences over (a) goals of development, (b) key resource constraints, (c) key institutional constraints, and (d) policy orientation. The pro-subsidy group believe that subsidies are necessary in order to encourage the scaling-up of industry. They perceive individual material welfare and state power as goals of development and identify low incomes as the major constraint on development. Their policy beliefs propose the reform of large financial institutions, through conditional lending, changes in legislation, training and technical assistance. The anti-subsidy group believe that subsidies reduce economic growth by distorting capital markets and inhibiting efficient financial intermediation. They perceive as the goal of development individual material welfare and freedom, and identify poor incentives to save and invest in financial intermediation as the main constraints on development. Their policy suggests the closure or privatization of large financial institutions, thus creating the conditions for more efficient financial intermediation through private-sector innovation (see Copestake, 1996, for a fuller discussion). The opposing positions of the pro-subsidy and anti-subsidy groups can and sometimes do exist within the same aid chain. For example, a donor agency may have a policy to promote private-sector innovation, while being opposed to the provision of subsidies on the grounds that they encourage inefficiency. This donor may be providing funds to an NGO who believes that unless small industries are subsidized they have very little chance of scaling-up their operations or even remaining viable. Such scenarios are likely to lead to tensions between the donor agency and the NGO, and may well result in a compromise policy which is less than effective in alleviating poverty.

These different agendas, as one might expect, create difficulties for NGOs and their donor agencies in agreeing a common goal or route of action. For example, 'donor co-financing of NGO credit programmes is of particular interest because it is an explicit, formalised credit subsidy, based upon voluntary agreement, and therefore requiring that differing donor/NGO views on credit subsidies are either reconciled or compromised' (Copestake, 1996, p. 28). The emphasis which the new policy agenda places on the importance of the private over the public sector, and on the close monitoring of the financial performance of lenders, may force NGOs to expect their credit programs to become fully self-sustainable. Unfortunately, the pressure on NGOs under close financial scrutiny to meet their self-sustainability targets are more likely to encourage them to offer subsidies to creditworthy individuals who, by definition, are already in receipt of funds from elsewhere, or, in the absence of these, to work with the most creditworthy of those who are not already receiving subsidies. Such a strategy is unlikely to benefit the poorest.

NGOS are, therefore, sometimes forced to compromise their anti-poverty focus in order to secure funding to ensure their own survival. Also, in response to bureaucratic pressures from their funding masters, NGOs are often accused of placing undue emphasis on starting new projects rather than managing and revitalizing old ones. 'Prestige and promotions accrue to staff who launch promising, big, new activities rather than to those who spend time on the more mundane tasks of everyday management' (Van de Walle, 1995, p. 239; see also Chapter 5, this volume).

In terms of the aid chain, the explicit agenda at each point may be that of poverty alleviation, but, as suggested above, more 'personal' agendas such as the organization's survival or an individual's desire for promotion can take precedence at any particular point in the chain. The inevitable result is that the more explicit agenda of poverty alleviation is compromised.

As we have already intimated, agendas are often influenced by political pressures or the more basic will to survive. Social pressures also play a part in the formulation and maintenance of agendas, particularly when the agenda-setting group or organization is uncertain to begin with, either in terms of the clarity of the outcome it is seeking or, more broadly, in terms of the ambiguity of the decision-making process itself. Aid committees may sometimes be reluctant to entertain new ideas, even in the face of negative feedback from the field (Porter *et al.*, 1991). Gow too has noted that aid organizations, such as NGOs, sometimes display a curiously 'dogmatic' tendency (1991, p. 3) to adhere to their initially preferred policy, even when there is factual evidence that fundamental mistakes are being made.

Converging to a set agenda

In social psychology, the term *convergence* describes the tendency for inherently ambiguous tasks to precipitate a certain amount of fairly rapid mutual concessions toward the mean or most common viewpoint (Sherif 1936, 1937). Convergence serves to reduce the anxiety of uncertainty, which has been described as inherent in international assignments in general (Mendenhall and Wiley, 1994) and to some aid projects in particular (Porter *et al.*, 1991). Sherif's work has not dated, and extends well beyond the laboratory. The same uncertainty-reducing convergence has, for instance, been demonstrated in organizational groups, for example, across 432 personnel managers, in a variety of European countries, who socially converged on a norm when setting new wage levels (Wilke and Van Knippenberg, 1990). NGOs, too, do not always feel wholly informed or confident about what precisely is expected of them, or indeed how their work will be evaluated (Thomas, 1996). Such uncertainty may also encourage convergence.

In Sherif's original demonstration of convergence (1936), people were shown the 'autokinetic [self-moving] effect'. This is simply a fixed pinpoint of light, which, when shown to people in a darkened room, appears to move in various directions and for varying distances. Sherif asked his participants to estimate the distance of each movement, first alone and then together in groups. He found that each individual would first converge on a personal standard, and that when put in groups these different standards would rapidly converge into one group norm. This norm would be an average of the views that individuals in the group held beforehand. In other words, the group norm, i.e. where members converged to, was in no way veridical. Moreover, these norms persisted for up to one year later. What we observe, therefore, is relative flexibility giving way to an arbitrarily fixed position in the face of inherent uncertainty.

In short, whenever an aid project (or decision-making process itself) lacks a clear solution, we would expect a certain movement toward convergence. Convergence can be viewed as a means of trying to cope with uncertainty. This would resonate with Porter *et al.*'s (1991) description of donor-funding agencies as being governed by 'control orientation' and with Hofstede and Bond's (1988) argument that Westerners are often relatively concerned with the certainty of 'one truth.'

Polarisation toward incompatible agendas

Where a group of people have a clearer and more informed stance at the outset, we might expect to observe a social process known as 'polarization' (Moscovici and Zavalloni, 1969). This is the tendency for an existing group norm, or position on an issue, to become amplified both during and after a group-consensus task, such as a committee meeting where decisions must be reached (Lamm and Myers, 1978). It is believed to be caused primarily by group members' pooling and thereby reinforcing each others' arguments for their initial position, especially when facilitated by a live discussion rather than passively read information ('informational' social influence). A secondary cause of polarization stems indirectly from subjective group pressures to 'tow the group line'. This more socially based, contributory process is especially likely when the issue at stake is charged with emotion (Isenberg, 1986). It is more directly attributable to the competitiveness of some individuals, who try to 'outdo' the group norm ('normative' social influence).

Group polarization is a robust social effect. It has been found on a variety of aid-related topics, such as ethical decisions and helping behaviour (Lamm and Myers, 1978), as well as in different cultural settings, including some developing countries (Smith and Bond, 1993). In respect of such traditional societies, however, it is important to remember that 'normative' social influence may not occur with the same intensity as in the more individualistic West, and that group polarization does not necessarily mean change. If the group is initially pro-stability, then that is the tendency that would polarize. In relation to international aid, the polarization literature provides us with an intelligible social mechanism by which decision-making groups, especially on the donor side, might become increasingly extreme in their initial leanings, whatever those might be. The pro-subsidy and anti-subsidy groups already described are a good example of this phenomenon.

Groupthink

Group polarization occurs most easily in newly forming groups or with relatively new issues (Smith and Bond, 1993). Over repeated meetings, each of these two scenarios is likely to develop into what has been termed 'groupthink' (Janis,

1982). With time, it may become increasingly unthinkable that one's group could be wrong in its chosen course of action. Set unanimity of opinion often reassures group members that they are right, and provides a powerful motivator in the form of certainty and cohesion (Festinger, 1950). As a result, internal dissension becomes less and less likely, even perhaps when aid projects clearly begin to leave the rails.

Project evaluations are all too often contaminated by groupthink, often to the extent that the objectivity necessary for an unbiased and fair evaluation is lost. As a result, projects may be mistakenly judged successful by evaluation teams and as a result may be perpetuated beyond their useful life. In general, project planners and project staff have to date tended to focus on immediate outputs rather than the ultimate, but often less tangible, indicators of effectiveness. As Berg (1993) points out, 'ultimate impact is forecast on the basis of projections about the cumulative effects of the immediate outputs. It is assumed, for example, that procedure manuals or computerisation will lead to greater efficiency in a beneficiary organisation' (p. 17).

In Janis's original study, he attributed the disastrous decisions of US Presidential Select Committees to groupthink, including the decision to invade the Bay of Pigs in Cuba, and to escalate the war in Vietnam. In each case, as errors of judgment about these situations became increasingly apparent, groupthink actually prevented corrective action from being taken. Instead, the group became ever more convinced of its own theories, an essentially delusional process that has been ascribed to aid decision-making bodies. The policies of the World Bank and IMF in relation to developing countries also need to be scrutinized in terms of groupthink on an international scale.

In sum, if we subject the processes involved when people work in groups to social-psychological analysis, we see that there are several grounds for believing that there could be coalescence and rigidity, rather than due consideration and flexibility. This is likely to be particularly so where there is poor communication between different aid chains or between different links in the same chain.

The communication process

Lack of coordination

Multilateral aid is accompanied by the baggage of many stakeholders which lead to multiple agendas. Conflicting agendas and multi-tiered administration and its ensuing bureaucracy can slow progress in development work. This negative image of multilateralism can prohibit the effectiveness of the multilateral agencies who play a key role in coordination, and encourage donors to return to bilateral aid and therefore less coordinated aid.

By the mid-1980s, Somalia had received more than twenty years of relatively intensive technical assistance in agricultural research and extension (Berg, 1993).

However, according to a UNDP/World Bank assessment (UNDP and World Bank, 1985) the discontinuity and fragmentation of assistance to agricultural research has meant that there has been little possibility of pursuing a consistent program of investigation over an extended period of time. Moreover, there is evidence that information gathered in the earlier stages of the research was not available to those involved in later stages.

However, not all of the blame for poor coordination can be attributed to the donors. Development projects which change the environment are complex by nature and touch on the concerns of many different government sectors. Personnel from each of these sectors are usually specialists, unfamiliar with the concerns of other sectors and untrained in communicating with them. There are usually few or no formal linkages between officers in different line ministries, for example agricultural ministries are likely to see health matters as falling solely within the realm of the Ministry of Health and vice versa. In a review of technical cooperation, Berg (1993) finds a recurrent theme, namely that 'technical cooperation projects are poorly planned, programmed and coordinated, largely because of a vacuum created by the absence of a strong central planning unit in government; the result is a multiplicity of uncoordinated donor-driven technical cooperation projects' (p. 8). This suggests that because of a lack of coordination at the lower levels of the aid chain (see Figure 3.1), the project agenda is decided by or driven from the upper link in it.

Both donors and recipients have had strong reservations about engaging seriously in coordination. On the donor side, there have been three main reasons: firstly, coordination is likely to impede the freedom with which donors pursue their political interests through their aid programs. For example, American aid 'sustained regimes in power with economic assistance long after they had proved their developmental incompetence beyond a reasonable doubt. The best example of this logic is probably Mobutu's Zaire, a particularly inept and brutal regime which received over US$1 billion in US assistance between 1962 and 1990, essentially because of its perceived usefulness to American intelligence operations in Central and Southern Africa' (Van de Walle, 1995, p. 240); secondly, donors know there are policy issues on which they will disagree; and thirdly coordination can be costly and time-consuming.

On the recipient side, there are concerns that coordination could result in unbearable pressures in terms of policy reform. Recipient countries have valued the freedom to play donors off against each other. Also it is not easy to sort out the differences among finance ministries whose primary concern is financial control, and other ministries who are primarily interested in spending on their own programs. Recipients would also be concerned about the administrative costs of coordination (Cassen, 1994). Poor coordination tends to allow for less accountability on the part of recipient governments, as any deficiencies in public services will at least in part be met by donors when the need becomes acute.

As an instance of poor coordination, Cassen (1994) provides the notorious example of 350 different expert missions that allegedly visited Burkina Faso in one year alone in the early 1980s. This resonates with the conclusions of a broad-ranging review of over 1,000 USAID projects, which found that ineffective organizational relationships was one of the principal factors differentiating failures from successes (Rondinelli, 1986).

At a consultation between WHO and NGOs working in Africa (WHO, 1996) the need for a coordinated multi-sectoral approach to tackling poverty was identified.

> Agencies, governments and NGOs need to use a holistic approach in coping with poverty. The intrinsic link between poverty and ill-health demands a coordinated approach to improve the health of the poor. This approach requires the involvement of health sector actors as well as actors from other sectors, such as agriculture, education and environment. Poverty and its negative impact on the health of populations cannot be tackled by the health sector alone.
>
> (WHO, 1996, p. 3)

The need to build partnerships among donor agencies, host governments, and NGOs was commonly agreed at this meeting. Building such partnerships would help to strengthen the links in the aid chain and in so doing would improve communication along it.

Interagency communication

The chain of communication portrayed in Figure 3.1 is similar to one of a series of networks that have been studied in social psychology (Leavitt, 1951; Shaw, 1964). Essentially, these 'communication networks' were varied in terms of the number of communication channels permitted between individual group members or links. At one extreme is the highly autocratic 'wheel', wherein one individual becomes the central hub for all communications. Close to this structure, in terms of number of channels, is a network known as 'the chain', which is structured exactly as in Figure 3.1. At the other extreme is the thoroughly egalitarian 'all-channel' structure, in which anybody can communicate with everybody.

The central finding of the communication-networks research has stood the test of time well. In Western settings, the more centralized the structure, and the more peripheral the group member's position in that structure, the less satisfied, on the average, that group member is likely to be. At the consultation between WHO and NGOs working in Africa (WHO, 1996) many of the NGOs expressed dissatisfaction at the quality of communication they had had with WHO. The meeting, which was held in Maynooth, Ireland, and was attended by representatives from more than fifty NGOs and aid agencies, made the following recommendations to improve the role of WHO:

- developing ... a project in selected sub-Saharan countries to promote an active *partnership* between Ministries of Health, NGOs and communities in targeting poverty and its adverse effects on health status;
- becoming more *accessible* through organising regular fora for NGOs to share experiences of implementation and practice; provide a platform for NGOs in WHO meetings on poverty and health, and in the renewal of Health for All;
- *collecting and disseminating* models of best practice and examples of successful programmes;
- integrating health and poverty statistics worldwide, and stratifying them for very poor/vulnerable groups, to *provide a guide* for NGO planning and action.

(WHO, 1996, pp. 7–8)

Clearly, the NGOs participating in this meeting recognized the need for better coordination in their efforts at improving the health status of the communities they work with. Further acknowledgment of the poor communications between links in the aid chain is evident in the NGOs' identification of the gap that exists between policy and action and the report's recommendation that:

NGOs ... provide an important link in the current gap between policies and actions to reduce poverty at country level. It is important that they develop this link position to keep the donor agencies informed, and work together with them to bridge the gap between policy and practice. This will ensure that planning and implementation are, as they should be, intrinsically linked.

(pp. 6–7)

Social identity

NGOs, unfortunately, are often in competition with each other. There is an intrinsic drive to succeed where others have failed, and it is easy to form the impression that if an individual NGO should happen upon the recipe for success there would be a temptation to keep it a closely guarded secret. According to Tajfel (1978), people are fundamentally motivated to be members of a group, and to have a group identity, in terms of which, by comparison to some comparable outgroup, they can feel justly proud. In the case of Figure 3.1, these outgroups might consist of the other links in the chain or rival NGOs (at the same level in the chain).

One of the emergent possible causes of polarization is knowledge or consciousness of some nearby outgroup. In the original demonstration that such knowledge in itself could be sufficient to produce polarization, French architecture students had to describe their self-concept (Doise, 1969). Whatever trends were evident in these descriptions became amplified in those groups that were also subtly reminded of the traits of a group in a nearby school of

54

architecture. Their concept of who they were had become polarized, in an effort to 'mark it out more clearly' (and positively so) from a rival group.

More recently Hogg *et al.* (1990) have shown how Australian students' initial tendency toward risky decisions became amplified just by being made aware that there was another group who tended toward caution; became more cautious when they realized that a nearby group was more risky; and 'converged' on themselves, i.e. consolidated their initial tendencies, when juxtaposed with two groups, one risky and the other cautious. Perhaps we should remember here that international aid projects, by their very nature, involve both risk and caution.

What these findings suggest in relation to the aid chain in Figure 3.1 is that comparing one's group (or organization/NGO) to others could be another contributing factor toward each of these group's developing their own polarized organizational culture. More than that, however, the findings suggest that each organization may attempt to stamp its own seal on the decision-making process, and on the project as a whole. Thus, Craig and Porter remark about the aid chain that 'each of these [NGO] organisations reframes and reconstitutes the project' in its own image (1994, p. 19).

Research on negotiation has also shown that the quality of relations between adjacent partners often has a substantial, and potentially distorting effect on decision quality (Pruitt, 1995). Decisions traveling further down the aid chain may be perceived increasingly as a constraint on collective autonomy, culminating in socio-cultural resistance analogous to that observed in the opening Case Study (see Chapter 1). In a Malawian context, Simukonda (1992) has commented on the failure of NGOs to consider necessary infrastructure capabilities in the host community, as well as the tendency for their failure to foster the community will necessary to sustain projects once the funding is withdrawn.

In the social sciences, communal myths and narratives generally appear to be gaining recognition as a motivating force in human behavior, and we explore this more fully in Chapter 8. For the moment, however, we might conjecture that the idea of 'democracy' itself, and its attendant forms of 'good governance', are themselves narratives. If so, the way they translate into action and reaction as they travel down an aid chain could easily generate intergroup tensions. In discussing Western expatriate managers' attitudes and preconceptions about African under-managers, Lane and Burgoyne (1988) argue that cultural myths and stereotypes often plague the communication process. Similarly, it was once held that 'managers throughout the world approve of participative and democratic management, without holding a corresponding belief in the capacity of people for initiative and leadership' (Clark and McCabe, 1970, p. 6). To the extent that this remains so, aid chains may actually bring groups to loggerheads about what is an appropriate 'script' for the 'democratization' of aid, or the 'civil society' that it may be designed to promote.

Participation

The rhetoric

In recent years there has been much debate on how to involve the aid beneficiaries in the aid process. Terms such as 'consultation', 'mobilization', 'participation' and 'empowerment' are commonplace in the literature. However, the reality of what happens on the ground does not always correspond with the rhetoric that any NGO realizes is necessary in order to obtain funding from donors. For example, in Malawi a UNICEF social-mobilization project, which aimed to mobilize communities to spread AIDS-prevention messages, did more to immobilize these communities because of a delay in releasing funds for such activities (McAuliffe, 1994a). Communities which might have organized such activities of their own accord, sat back and waited for UNICEF funds that were tied up in the bureaucratic system of project administration.

In the 1990s, the IMF and the World Bank have shown greater receptivity to the involvement of employers' and workers' organizations in the structural-adjustment process. At the IMF, the managing director has stated that governments are more likely to succeed with their economic strategies if they are supported by a popular consensus (IMF Report, 1992). An article in a World Bank publication has also suggested that 'the crucial element in gaining worker's co-operation is the belief that their sacrifice will contribute to general gains and that those gains will be distributed fairly' and that 'co-operation might be encouraged by mechanisms to give labour fuller access to decision-making circles, while ways are sought to broaden labour leaders grasp of national problems' (Nelson, 1991, p. 53).

As we have argued, SAPs have, to date, concentrated on reductions in public expenditure, more in the social-services area than in any other (Rahman, 1991). This has resulted in major sacrifices being requested of those who are poor, both economically and in terms of their health. These people need a voice, and it has been suggested that 'timely, meaningful consultation with representatives of the major social groups can make the difference between a policy that is resisted and sabotaged and one that has been shaped in part by the people affected by it' (Trebilcock, 1996, p. 7).

Many development theorists and practitioners now see participation and empowerment as critical to the success of development projects. Support for the participatory approach is in vogue with both multinational and private-sector agencies. The UNDP's 1993 Development Report describes this as 'a revolution in our thinking . . . that makes people's participation the central objective in all parts of life' (p. 8). And the World Bank states its commitment 'to support government efforts to promote a more enabling environment for participatory development within client countries' (World Bank, 1994).

However, these objectives run counter to the way in which these organizations currently make decisions in their large hierarchical bureaucracies and with little consultation of the people who might be affected by such decisions. Thus, as Brett (1996) states, 'participation might dominate development discourse, but hierarchy remains its dominant practice' (p. 6).

System constraints

Even when consultation does take place, it does not always lead to changes which benefit the poor. In the case of Zambia, for example, although the move in 1991 from one-party rule to a democracy, headed by a former trade-union official, has meant much closer involvement of the social partners in structural-adjustment decisions, it has not meant freedom from criticism, or from conflict over unpopular government policies aimed at rectifying some of the major distortions in the economy (ICFTU, 1992).

Some Western-based research suggests that a very negative influence will result from under-fulfilled promises of 'participation'. Baldwin et al. (1991) used three methods of implementing a staff-development project. One group was simply told what training they would receive; another was allowed to choose the (same) package; a third group was led to believe that it would exercise a degree of choice but did not, in the final analysis, participate. The latter group was subsequently the least motivated of all, and learned less than either of the other two groups.

In systems-theory terms, creating an expectation that democracy will prevail and then failing to meet that expectation could be conceived of as 'a fix that fails' (see Chapter 1). Here, from Figure 3.2, a 'band-aid' ('sticking plaster') solution, after a delay, creates more problems than it solves. Creating the expectation of full participation and consultation may initially 'smooth the way', but with time, the in-built constraints of the aid chain will begin to frustrate that expectation. As Senge puts it, such unfulfilled and bland promises 'will eventually drive out real vision, leaving only hollow "vision statements", good ideas that are never taken to heart' (1992, p. 231).

In a Kenyan context, Porter et al. (1991) have vividly described how international aid does not always keep its promises for 'participation'. This concurs with observations from within Malawi, where Simukonda (1992) has asserted that NGOs often give the false impression that the community will be allowed to set its own aid agenda. Instead, Simukonda argues, development projects are more likely to reflect fixed ideas developed by donor groups, and the desire to test their universality. With regard to technical-assistance projects to train the trainers, Fox (1994) describes how promises of participation are often perceived as insincere, eventually giving rise to feelings of betrayal among the so-called 'participants'.

Why then has it proved so difficult for so many aid workers to put their beliefs and ideas about participation into practice? The desire to engage in meaningful consultation and form real partnerships with the communities the NGOs work with does exist. However, in practice, the constraints which exist in the bureaucracy of the international aid system make it very difficult for individual

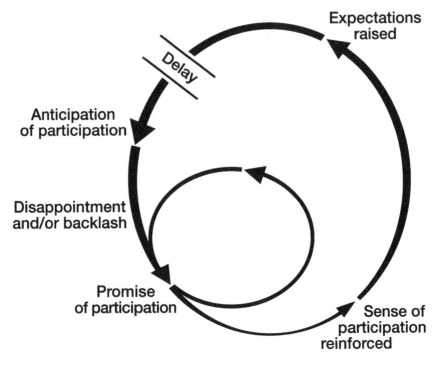

Figure 3.2 A participation fix that fails

aid workers to operate in the way they would like. These constraints range from expectations about timeframes for projects, what constitutes success or failure, and how aid workers should operate, to what type of lifestyle they should have in the developing country. These expectations serve only to imprison the aid worker in an outdated and ineffective *modus operandi*. In essence, it is the social structure of the aid organization that prohibits aid workers from enacting the widely agreed rhetoric about participation. It is the aid chain itself which makes it (only) rhetoric.

Types of power

Power and control play a major part in creating barriers to real and meaningful participation. Power can be defined as 'the capacity to alter the actions of others' (Kelman 1974; Thibaut and Kelly 1959). There are several different types of power available to people to bring about this alteration. The best and most enduring taxonomy of power types was devised by French and Raven (1959; see also, Handy, 1985).

Reward and *coercive* power are based on the capacity to wield any kind of carrot or stick, respectively, and if used excessively tend to produce superficial (purely behavioral) compliance, requiring constant surveillance, rather than inner acceptance and commitment. These are not normally the kinds of power that would be publicly associated with international aid, at least by donors, but possible examples might include aid-tying and 'coercive democracy'.

Referent power is based on liking, and this tends to have a somewhat more profound, 'identification', effect psychologically, but lasts only for as long as the liking relationship itself lasts. In international aid projects, heavy dependence on 'charismatic' leadership from donors abroad often creates problems of its own, in particular during the all-important transition between reliance on visiting expatriate referent power and longer-term, sustainable development once the project itself is over and the expatriate has left for his or her home (Kaul, 1989).

Legitimate power is based purely on holding some recognized position within the group, and may produce either public or privately accepted influence, depending on the level of recognition. Traditionally, communities in many developing countries tend to place comparative weight on legitimate power, which is often vested in, and wielded by, elders and other senior community figures.

Finally, *expert* power is based on demonstrating competence in a given area, and is closely connected to the idea of credibility. This form of power is totally non-coercive, and for that reason tends to produce internalized (inwardly accepted) influence, which can be the most durable and sustained form of influence. In international aid projects, credibility is often vital for exercising influence in the field (for example, Salmen, 1995), and has long been recognized as such (for example, Rogers, 1971).

Many NGOs and aid agencies maintain a position of power relative to the communities they work with. Traditionally, aid and development work, particularly where there is a strong religious involvement, has been paternalistic in nature. As a parent attempts to control the behavior of his/her child, so too the aid agencies have attempted to control the behavior of the people they sought to help. It has been difficult for aid workers to shed this mantle and many continue to exercise control through the use of the expert power invested in them by the communities they work with (by virtue of the fact that the communities believe that the aid workers have the knowledge and expertise to solve their problems), and the use of reward and coercive power (they have control over resources and they decide the criteria for the distribution of those resources). Although the use of expert power has the ability to produce change in people's personal beliefs and feelings, the use of reward and coercive power tends to produce compliance rather than real change, and people will only comply when there is surveillance (Gergen and Gergen, 1981). In other words people are only changing because someone is 'waving a carrot or a stick' and as soon as it is withdrawn they will revert back to the 'old ways'. Change which is brought about in this manner is therefore unlikely to be sustainable.

Any serious effort at empowerment must first acknowledge that in order to give power to the communities themselves, it is necessary for NGOs and aid agencies to relinquish some of their power. However, many NGOs find it impossible to give communities the power to decide their own agenda. There are several reasons for this: firstly, it takes considerable courage to approach the community with a blank slate, especially if there are expectations that the NGO will provide all the answers; secondly, NGOs may find it very difficult to persuade their funders to fund such an unstructured and risky venture; and thirdly, the timeframes imposed by donors do not allow for lengthy problem exploration and agenda setting.

In an analysis of the social-psychological effects of power, Kipnis (1977) describes the steps by which the powerful become corrupted as follows:

1 *Access to means of power increases the probability that power will be used.* If reward and coercive power are available (as they usually are in aid work) these are far more likely to be used than persuasive power, thus making it less likely that results will be sustained.

2 *The more power used, the more likely the power-holder will believe that he or she controls the target's (i.e. the person the powerholder is exercising power over) actions.* Aid-project workers who exercise more expert, reward, or coercive power are less likely to perceive the host community as being self-motivated.

3 *As the power-holder comes to take credit for the target's actions, the target may seem to be less worthy.* If aid-project workers believe they have brought about the change, they will tend to devalue the input of the community.

4 *As the target's worth is decreased, his or her social distance from the power-holder increases.* The less aid-project workers value the input of the communities they work with, the more they will distance themselves from the community and the less likely they are to consult them.

5 *Access to and use of power may elevate the self-esteem of the powerful.* Aid agencies may use their position of power relative to the host communities to boost their self-esteem. This can lead the host community to further devalue themselves.

The following case study describes a method of reducing the power differentials between aid project workers and the communities they work with. This 'leveling' technique was used effectively to give control of the project to the target group themselves. NGOs and aid agencies need to understand that, paradoxically, they can only gain 'ownership' and claim credit for projects if they are first willing to relinquish control.

Case Study – Mahruh Resource Management Project, Egypt

In 1990, the government of Egypt asked the World Bank to help identify ways to improve agriculture in the Mahruh Governorate, particularly for poor, isolated farmers. The Governorate is located in the western desert bordering on Libya in the west, the Mediterranean Sea in the north, the

Sahara desert in the south, and the Nile Delta in the East. The population of the area is approximately 0.25 million, of whom 85 percent are Bedouins, organized as a traditional Bedouin society, modified during the last decade as they moved from a nomadic to a more sedentary lifestyle. Despite attempts by the government to bring the Bedouin into the mainstream Egyptian society, they remain an isolated tribal society with their leadership performing many administrative and judicial functions.

The World Bank concluded from an initial identification mission that a traditional livestock programme would not adequately meet the needs of these people. Instead, they decided that the Bedouin needed a project to assist them in improving resource management, focusing specifically on collecting and storing rainwater for irrigation purposes. The Bank decided to adopt a novel approach to the project and instead of sending in a team of external experts, the local people did the work of identification and preparation for the project. They first established a local task force consisting of people from the central government, local government, local institutions and the Bedouin community. This task force, assisted by a UK consulting firm, used the Participatory Rural Assessment technique to develop participatory maps, conduct transect walks, develop social and historical profiles and seasonal calendars. As part of this process they used 'leveling techniques' i.e. techniques to reduce the power differentials between the various stakeholders and so level the playing field. One example describes how outsiders watched respectfully as the Bedouins drew maps on the ground. Another involved role reversal, with the Bedouins leading outsiders on transect walks i.e. walks to the periphery of an area to gain an understanding of the spatial differences in terms of land-use, vegetation, agricultural practices, etc.

The project preparation phase for this project took a whole year. However, the participatory approach added one important value which has been described by the task manager (Bachir Souhlal) as follows: 'the Bedouins realized that we were not attempting to use them, as had been their previous experience with outside authorities. This opened the way for trust to occur, and the trust became mutual before long. I can not imagine that occurring when a group of external experts rush in and out gathering facts and making judgments and quick recommendations' (p. 44). Souhlal succinctly sums up the Bank's experience of this project 'Through participation, we lost "control" of the project, and in so doing gained ownership and sustainability, precious things in our business' (p. 44).

Adapted from World Bank Participation Sourcebook (1995)

A task focus vs. enabling meaningful participation

Brett (1996) argues that complex activities can only be jointly managed where the members have the expertise and information required to sustain them. Cooperatives require higher levels of skill and social awareness than privately run firms. This is not often acknowledged by donor agencies and NGOs. The poorest and least educated are the least likely to possess these skills, yet many NGOs assume that they need only a few bags of cement and some encouragement to build a hospital or a community centre. They are task- rather than process-focused, which is understandable in a context where success is often measured by what one gets done rather than how one does it.

The consequences of failure to acknowledge the need for a lengthy process of education and training, in order to run a successful cooperative, are demonstrated by the experiences of one NGO's income-generating project in Malaŵi. The NGO attempted to establish a cooperative among women who did not have a steady source of income. The NGO decided on an income-generating project in egg production, where each woman was provided with a couple of laying hens. The objective was to encourage the women to work together to find markets for the eggs produced by these hens. However, the project failed when many of these women took the decision to kill the hens to feed their families now rather than wait for the future income from the sale of the eggs. Clearly there was a mismatch between these women's priorities and the NGO's expectations of them.

One of the factors which influences the decision to change is having the skills and self-efficacy to bring about that change. McAuliffe (1996) in a study of the barriers to adopting safe-sex behaviors in Malaŵi illustrates that perceived self-efficacy may be strongly influenced by norms and cultural values. This is particularly true in the context of sexual relationships. Women believe that the male, as head of the family, is the decision-maker, and therefore if condoms are to be introduced in the relationship, the male has to make this decision. In such cultures, where there is an obvious power imbalance between the sexes, enactment of behavior change proves extremely difficult. 'Effective health education may produce a commitment to behavior change. However, if this commitment is to lead to enactment, the powerless position of women needs to be directly addressed as part of any health education programme' (McAuliffe, 1996, p. 384).

In order to engage in meaningful consultation and participation, representatives of vulnerable groups such as poor women and rural populations must be represented. There is of course a difficulty in assessing the legitimacy of representatives of such groups, but this is not sufficient reason to accept token representation. Governments and donor agencies alike also have a duty to ensure that any such representatives are sufficiently informed to allow them to contribute meaningfully to the decision-making process. If the individuals being asked to participate are not sufficiently informed and knowledgeable regarding the philosophy and objectives of both the participatory approach and the project, then it is likely that their judgments and decision-making capacity will be constrained by bounded rationality (or ignorance). It is equally likely that their commitment to the project/cooperative may be influenced by self-interest and opportunism (see Chapter 11). Cooperatives usually operate according to an egalitarian ideology which assumes similar inputs, in terms of skill and commitment, and an equal distribution of the outputs or rewards. However, as Brett argues, this scenario rarely exists in development projects. Instead, he argues

> even in very simple activities carried out by PVOs (participatory voluntary organisations) in poor countries, a few activists will always make a much larger contribution than the majority of the members who are, therefore, free-riding at their expense. These activists will probably lose their motivation unless their commitment to the project is overwhelmingly strong, they can be rewarded through prestigious positions, or they can be paid some form of fee. Where this is not so, they will commonly adopt corrupt practices, which soon delegitimise the enterprise.
>
> (1996, p. 16)

In a study on tripartite cooperation and consultation, Hernandez Alvarez (1994) put forward a set of necessary conditions to improve the chances of success of such consultations, which can be summarized as follows:

- a climate of political stability and respect for rights that permit the exercise of freedom of association;
- a genuine commitment on the part of each participant to engage in social dialogue toward consensus, subordinating narrow interests to those of the community as a whole;
- provision of sufficient information to the social partners and their capacity to analyze it to reach rational conclusions;
- a degree of strength of the participants in the consultative arrangements that ensures that they are sufficiently representative and that they have enough influence over their members to make a commitment that can be kept.

In addition to the above, we would add the importance of timely consultation. Many donors consult only after the policies have been decided, as in the situation where one of the authors was asked by the World Bank (but refused) to conduct a rapid rural appraisal to determine public reaction to their new family-planning policies, one month before the launch! (This was to be the first consultation with those that might be affected by the policy.) Consultation must take place at all stages of the process, from the information-gathering, to the setting of priorities, to the articulation of policy and its implementation, to the evaluation of its impact. To date, the problem has been that if representatives become involved in SAPs, 'it is at a very late stage, and certainly after a definition of relevant information has been made and once the priorities have been set. Often, these representatives are called in only in an attempt to find a palliative to social unrest that could have been avoided had they been involved from the beginning' (Trebilcock, 1996, p. 13).

Participation in development projects is therefore about more than giving people a voice. It requires a commitment to ensuring that those involved acquire the necessary knowledge and skills to participate in an active and meaningful way. It must also be borne in mind that individuals participate in cooperatives not because of altruistic motives alone but on the expectation of reciprocity (see Chapter 11). In order for the cooperative to work effectively, others must meet their side of the bargain. Ensuring that this occurs is difficult in the absence of sanctions, and Brett (1996) suggests that the survival of organizations requires that reciprocation is an institutional obligation with the threat of negative sanctions should it fail. This would also help to reduce the type of opportunism that commonly occurred in African marketing cooperatives, where elected leaders exploited their members, who were handicapped by their lack of information, and also by their position relative to, and their dependence on, the leader (who was often the tribal chief). Thus, and to presage the next chapter, it can be argued that true participation that results in effective decision-making requires a cooperative with a balance between participation and hierarchy, and not on one or the other.

Part 2

DONORS ABROAD

4

LEADERSHIP, PARTICIPATION, AND CULTURE

> In international development, as in other policy domains, exhortations to take culture into account are plentiful, but the know-how is scarce.
>
> (Klitgard, 1995, p. 138)

A knotty issue

The very idea of international 'aid', and its ethos of 'helping people to help themselves', suggests the notion of leadership. This is very commonly defined as influencing others; which in international aid-project terms becomes influencing development *through* others (Rondinelli, 1986), or facilitation. As Mohan Kaul puts it, 'Good *leadership* is essential to project success' (1989, p. 17, emphasis added). Hence, in the opening case study (see Chapter 1), a managerial 'expert' was expatriated to Tanzania, to assist a mining organization to 'develop itself', through his supposed facilitatory 'leadership'.

Thomas warns us, however, that *good* leadership does not equal 'guidance' from those more 'advanced' or culturally 'superior' (1996, p. 98). An important and long-standing question for leadership theory, research, and practice is whether (or rather when) a manager should focus just on the task at hand, or on respecting social relations (with)in the work group (Mayo, 1949; Taylor, 1912). Leadership often boils down to one basic question, namely, what degree of worker participation is appropriate? In aid projects, this can range from minimal local consultation to local people having full control over key decisions at the design, construction, implementation, and assessment stages (Squire, 1995, p. 34; Salmen, 1995, p. 150).

The Tanzanian example was not an isolated incident. It *generally* makes a difference whether one listens to intended 'beneficiaries'. At the assessment stage for instance, Salmen (1995) has described how he listened to: Bolivian urban renters whose tax-liability voice had been obscured by more prominent project beneficiaries; Thai core-housing beneficiaries, who did not move in to their new housing because they could not afford to furnish it, and feared losing face; rural parents in Mali, who did not send their children to an aid-funded (but distant) school, because they could not afford to provide them with 'packed lunches'; and people from Lesotho, whose traditional healers were preferred over new, 'modern' health workers, largely because the former always had curative medicines whereas the latter often had only (prevention-focused) words, combined with haughtiness. As you can well imagine, each of these complaints (and constraints) was relatively easily rectified, once the project leader (Salmen) allowed himself to listen to, and thereby be influenced by, the intended aid beneficiaries.

The issue is not as straightforward as these examples imply, however. Participation and a relations-orientated style of leadership are not always appropriate. As Salmen himself points out, some NGOs (often those affiliated with religious organizations) are more paternalistic than participatory, and yet manage their projects very successfully; meanwhile others – despite their overtly 'participatory' style – may fail miserably. Such failures, especially perhaps in a time of free and 'democratic' markets, when 'everyone can have their say', begin to suggest that the issue of participation is not quite as straightforward as it seems to be. Thus, across Senegalese and Kenyan contexts, for instance, completely opposite leadership styles were found to be appropriate (Rondinelli, 1986).

Group values appear to differ systematically (and to that extent predictably) between various (sub)cultural settings, and this fact is gaining increased recognition in aid work (for example, Klitgard, 1995). In *cross*-cultural assignments however, such as one finds in Technical Cooperation (TC) projects, the situation becomes far more complex. In this chapter, we explore how known differences between cultural groups might be applied to improve decisions about when 'participation' is, and is *not* appropriate.

As a case in point, we consider the technique of goal-setting (Locke, 1968). This has often been translated, in the organizational literature (for example, Muchinsky, 1993), as Management by Objectives, or MbO (Drucker, 1988). In theory, as originally expounded, MbO was a highly 'participative' technique, but in practice, it may have turned out to be pseudo-participatory (Halpern and Osofsky, 1990). But the idea of employees 'having a say' in how they work has had, and continues to have, a huge influence on Western management practice (Porteous, 1997; Statt, 1994). Today, for instance, it may be more appropriate to talk about 'performance management', and related techniques such as all-round, 360-degree feedback (Kettley, 1997; Tornow, 1993).

The essential point then is that each in turn has epitomized what participatory management is supposed to be about. For all that, however, we analyze how, why, and when each could be wholly inappropriate for aid work, in particular, for expatriate aid-project assignments. In our analysis, we will rely heavily on the taxonomy of power described in Chapter 3.

Trait theory

Perhaps the most intuitively obvious place to begin our discussion of leadership is with 'personality'. That is where studies of leadership did in fact begin, by attempting to construct a 'great man', trait-based theory of leadership. This was in accordance with the 'common-sense' idea that good leaders are born and not made. From the early 1900s to the 1940s, such theorists were measuring which personality traits differentiated effective from ineffective leaders, and leaders from followers. However, as the number of situations investigated grew, so too did the number of personality traits comprising the elusive 'great man'.

With the benefit of hindsight, that was not particularly surprising. Even within those comparatively homogeneous Western settings where the research was conducted, the qualities that make a good bank manager are liable to differ substantially from those that make a good general, or team captain, or captain of industry. Even abilities, notably 'intelligence', do not bear a clear link to successful leadership. A certain, minimum level of intelligence may be necessary to be an effective leader, but leaders can easily become 'too smart' for their followers (Gibb, 1969). They may actually appear more acceptably 'human', if they make the odd (but relatively inconsequential) mistake (for example, Aronson *et al.*, 1966).

Emergent leadership

A more productive emphasis has been laid on the process by which a leader gains his or her influential status. This approach has concentrated on the accumulation of both liking and trust, that is, referent and expert power. These are earned, according to Hollander (1964), by conforming in a Machiavellian-like way (but showing respect) to a group's norms, as well as by making competent judgments on behalf of that group.

Hollander has been able to show, for example, that equally expert leaders exert more influence, in the long run, when they first conform to group norms and only later start challenging them (and trying to exert influence), rather than challenging them from the outset. In order to maximize influence, leaders should earn both expert *and* referent power, which together constitute the essence of 'credibility'. In each aspect, they must be perceived as reciprocally *serving* their followers (Jones, 1990), a point made over 2,000 years ago by Confucius (Chinese Culture Connection, 1987), and more recently in regard to Asian economies (Hofstede and Bond, 1988). In the aid setting, Rogers (1971) has described how local community leaders who were seen to be 'bowing and scraping' too much to aid agencies soon lost their positions of influence and trust within their own community. More latterly, Salmen (1995) has reiterated the importance of credibility, in particular, for aid-project extension agents.

Styles and roles

A common foundation

Implicit in Hollander's work on leadership is the idea that a task focus and a relations focus often complement one another, suggesting that the optimum leader needs to wield more than one sense of power. Complementing and developing this suggestion even further is work on leadership 'styles' and leadership 'roles'. These two schools have studied the closely related topics of characteristic leadership tactics ('styles') and the needs of the group to be guided and nurtured ('roles'). The advent of these studies also marked the return of personality, although in a limited, comparatively malleable and two-dimensional sense (task vs. relations). Nonetheless, the two dimensions remain widely accepted to this day.

The common foundation of these approaches was arguably laid in a study by Kurt Lewin and colleagues, of boy's clubs and adult supervisors (Lewin, *et al.*, 1939). Lewin *et al.* found that task specialism, being based on reward and coercive power, was only moderately more productive than a more participative (and relations-focused) style, which in turn was based on referent, expert, and – possibly – legitimate power. Moreover, quality of production was higher under a participative leader, where in turn productivity also became higher whenever work was not closely supervised. The latter two findings may have been partly due to the fact that task specialism often generated an undertow of hostility toward the supervisor.

Since the original study, broadly similar findings have been made in industry (for example, Coch and French, 1948), in work settings in the Far East (for example, Misumi and Peterson, 1985), and, more recently, in some developing countries (for example, Jin, 1993). Importantly, these studies have not simply 'imposed' the investigators' own definitions of 'styles' or 'roles' on to the respondents. Often, the descriptions have been arrived at only after an extensive process of consultation with, or observation of, actual leaders and followers.

Styles

Let us take first of all the 'styles' approach, adopted during the 1940s and 1950s, and break it down into task- and relations-orientated styles. This basic inductive finding seems to hold well cross-culturally (Smith and Bond, 1993). In Papua New Guinea, for example, the traditional 'Big Pela' ('big man') style juxtaposes with the group-consensus tradition (Savery and Swain, 1985). In Ugandan educational management, there has been discussion of more 'masculine' (task) and 'feminine' (relations) leadership styles (Brown and Ralph, 1996). And in expatriate work generally, there is widespread recognition of both styles (Giacalone and Beard, 1994).

Moreover, although most individual leaders would tend toward one or the other style, the most effective leaders of all manage somehow to combine *both* relative 'extremes' of the continuum (Blake and Mouton, 1978), a finding that also holds in the West (for example, Dooley, 1996), as well as in the Far East (for example, Misumi and Peterson, 1985), including developing countries like India (Sinha, 1990). In the aid-project literature, these *three* possibilities (i.e., task, relations, and what we might term 'task-plus-relations') do seem to have been envisaged, but seldom in terms of a blend of high task and high relations focus forming the optimal balance. As Thomas puts it, 'to combine a study of development and management at all is rather unusual' (1996, p. 105).

Roles

The 'roles' approach seems to have been more clearly neglected (although see Belbin, 1981, and Chapter 8, this volume). According to this approach, groups need to be led toward (1) a solution and also (2) good interpersonal relations, needs which create naturally and spontaneously occurring 'scripts', in which there are two basic roles to be filled. This was originally observed in the West by Robert Bales (1955). He used the naturalistic, non-participant technique of watching and recording spontaneous interactions in initially leaderless groups, as they met over several occasions. Bales's observations led him to conclude that task specialists often emerge naturally (these were initially leaderless and structureless assemblages of people, given management case problems to solve), but soon begin to grate with the rest of the group. This grating then creates a second 'role', namely for a person who is more socially rather than task skilled, to 'keep the group together', and to meet their more-or-less 'natural', perhaps face-related needs.

Bales's work is relevant to aid projects and aid project settings. One of its key implications is that the leadership function can be performed by two individuals – especially since, as we have seen, people rarely manage to combine both task and relations skills. In a contemporary aid context, Salmen (1995) describes how two NGOs, one task orientated and paternalistic and the other relations orientated and participative in style, came together to combine their different and complementary leadership skills, and ultimately improved aid-project outcome. In Uganda, in an educational-management context, Brown and Ralph (1996) describe how an optimal balance of leadership styles is achieved by ensuring that heads and their deputies always fulfill complementary roles.

Contingencies

Such coordination, of course, is not always possible, and one response to this, in the West, has been to develop, during the 1960s and 1970s, again by inductive methods, a taxonomy of which style will work best under what conditions (Fiedler, 1978). The key elements in Fiedler's influential taxonomy are: preferred leadership style (task/relations, or a preference of wielding reward and coercive power versus referent power); position power (holding high versus low legitimate power, complemented – according to Fiedler – with reward and coercive power); task structure (clear versus unclear, as in building a washing machine to a blueprint versus deciding on welfare/aid policy); and leader–member relations (holding referent, and perhaps expert power, to a high or low degree).

If we consider the three (bipolar) situational factors in Fiedler's taxonomy (i.e.,

Table 4.1 Fiedler's contingency approach

Position power	Weak				Strong			
Task structure	Unstructured		Structured		Unstructured		Structured	
Leader/group relations	Bad	Good	Bad	Good	Bad	Good	Bad	Good
Best style	(1) Task	(2) Relations	(3) Relations	(4) Task	(5) Task	(6) Task	(7) Relations	(8) Task

Source: adapted from Fiedler (1978, p. 74).

position power, task structure, and leader–member relations), there are $2 \times 2 \times 2 = 8$ possible combinations or contingencies in which leaders operate. Fiedler then asks, which of the two styles, task or relations, is most often the most productive? Over many years, and in literally hundreds of studies involving thousands of people, across a great variety of organizations (but still mainly in the West), Fiedler has built up a picture of which style tends to work best in each contingency, and the information contained in Table 4.1 has proved itself fairly reliable (there). As one measure of this, a training package (called 'Leadermatch') has been developed, to help managers create a better 'fit' (Hesketh and Gardner, 1993) between their preferred style and the work environment with its ever-changing contingencies.

The beauty and major strength of Fiedler's model is that it brings together, under one roof as it were, all the elements of leadership that are important in the leadership (and aid) literature. Under leadership style, we have task and relations focus, embracing characteristic 'styles' or 'roles', plus reward and coercive power. Position power seems very close to legitimate power, which we have seen is often very important in traditional settings. Leader–member relations clearly pertain to referent, and possibly expert power, thereby incorporating the concepts developed through studies of emergent leadership. And finally, as we have argued in Chapter 3, (unclear) task structure seems to pertain to the inherently uncertain nature of international aid.

The broad notion of (local) leadership contingency is thus clearly germane to aid work. As Thomas puts it, for disaster relief it is probably appropriate to think in terms of 'command and control' mode, while giving advice to small businesses might be functioning in the mode 'empowerment and enabling' (1996, p. 100). Yet, with the partial exception of the former Soviet Union (where the overall notion of leadership being contingent has received some direct support, see Zhurvalev and Shorokhova, 1984), we do not (yet?) have anything like Fiedler's contingency table for developing countries, and certainly not for international aid assignments. And because the original model is not deductive, but rather inductive (i.e., it works backwards from actual data rather than forwards from theory), *it is impossible to guess whether it would transfer to other settings.*

For example, legitimate power is generally characteristic of Japanese work situations as well as those found in many developing countries (Misumi and Peterson, 1985). Misumi suggests that Japanese bosses tend to have very high legitimate power, and so may not even need to think of, or deal with, quite so many contingencies as in the West. In addition, the emphasis on quality (for example, through quality circles) seems to render 'M' (for 'Maintenance', i.e., a relations focus) more often superior to task-centered 'P', for 'Performance' (akin to 'Task' from Table 4.1), though this is not so in Fiedler's table. Moreover, and consistent with Sinha's (1990) Indian finding that relationship 'nurturance' is often best made contingent on successful task performance *itself,* Misumi reports that the comparatively rare top leaders are often just moderately performance focused on the job, and prefer to punctuate its completion with M-type activities.

73

Overall, therefore, it seems that Fiedler's notion of contingencies is important for understanding leadership. Included among these contingencies, however, should be *cultural* differences. Here, the taxonomy fails to provide us with any framework, or dimensions, within which cultural contexts could be safely assumed to vary systematically. In relation to international aid projects, *culture is the 'Achilles Heel' of Fiedler's inductive model*. It is therefore fortunate that cross-cultural psychologists have investigated this very question, and particularly so since they have often done so from an organizational point of view.

Cultural differences

The best-known figure here is Geert Hofstede (1980a), who, during the late 1960s and early 1970s, analyzed survey data from over 116,000 sales and service employees of the multinational corporation International Business Machines (IBM, pseudonym 'Hermes'). Hofstede was interested in work motivation in general, and in work values in particular. Included in his sample were samples from forty nations from around the world, including some so-called 'developing countries'.

Crucially, the level of analysis was cultural, not individual. This meant that Hofstede averaged item scores, measuring work values, over individuals within each country. He argued that this procedure would result in conservative (and reliable) measures of cultural differences, because IBM's corporate culture would, if anything, act to match people and therefore to *suppress* socio-cultural variation. Hofstede then factor analyzed the country mean scores ($N = 40$), effectively identifying the key directions, or senses, in which the cultures in question differed on the range of work values sampled by the test items. What emerged from this mammoth analysis were four major factors.

The first and major factor to emerge was individualism–collectivism (see Chapter 2). One example of a typical item loading on this factor is, 'Competition among employees usually does more harm than good', with which respondents were required to express their level of (dis)agreement (a collectivistic response would be to agree). The United States scored most individualistically of all on this factor, with two developing countries, Ecuador and Venezuela, scoring most collectivistically.

Another factor seems to have captured or reflected material and personal acquisitiveness. Examples of items include: 'It is important to have an opportunity for high earnings', and 'It is important to get the recognition I deserve'. This is probably the most controversial of Hofstede's factors, partly because he labeled it 'masculinity–femininity', on the grounds that men are universally (and by implication naturally) more assertive than women! (for example, Japan was most acquisitive, Sweden least).

A third factor was termed, by Hofstede, 'uncertainty avoidance', a term that was meant to denote a desire for rules, regulations, and certainty in life. An exemplary item loading on this factor would be, 'Company rules should not be broken even if the employee thinks it in the company's best interest'. In our view, the label is misleading, insofar as it suggests a generalized uncertainty avoidance, whereas the items are more workplace-focused. In developing countries, as we will argue in Chapter 10, one can be uncertainty avoiding in the workplace (because of economic job insecurity, for instance) but extremely tolerant outside of work (where pluralistic attitudes and practices would be, and are, extremely adaptive). Whatever the case, in Hofstede's survey, Greece and Portugal scored highest on this uncertainty avoiding factor, with Singapore (arguably, a very rule-based *society*) coming last.

The remaining factor Hofstede termed 'power distance', which means that hierarchy is generally preferred over egality. An exemplary item would be, 'Power-holders are "entitled" to privileges'. As such, power distance clearly over-laps with French and Raven's (1959) concept of 'legitimate' power. Indeed, 'developing' countries at the time, like Malaysia, Panama, Guatemala, and the Philippines, scored highest on this dimension. Last on the list were Austria and Israel.

Since Hofstede's seminal work, individualism–collectivism and power distance have by far captured the most attention, and we believe that these dimensions are the most valid and useful for our purposes. Somewhat counterintuitively, these dimensions were also negatively correlated with one another, which means that the most power-distant nations were also the most collectivistic. Put another way, and in terms of our cultural maps in Figure 1.1, the Western countries (high individualism, low power distance) were completely opposite to the developing ones (collectivist and high on power distance). Following Rudyard Kipling, some writers (for example, Kao and Ng, 1997) have referred to this, at least in the past tense, as the great 'cultural divide'.

Moreoever, these differing configurations suggest radically different concepts of participation. In effect, high collectivism is best envisaged here as a series of concentric, but ever-decreasing circles, arranged in a tall and hierarchical 'cone', rather than some kind of two-dimensional disk wherein each person is supposedly more 'equal'. We have tried to picture such a cone in Figure 4.1. More recent analyses have been suggesting that Hofstede's 'averaging' across entire cultural groups has missed certain more 'horizontal' (egalitarian) forms of collectivism, such as Australian 'mateship' (see Chapter 9, this volume) and Swedish 'Royal Envy' (Daun, 1991; Schneider, 1988). The important point for the moment, however, is that more 'democratic' forms of participation may not be culturally appropriate in less egalitarian, more power-distant settings, i.e., in some developing countries.

Are Hofstede's findings *reliable* though? After all, he surveyed just one (per-haps very particular) multinational, at one particular era of history some time ago! Other studies have in fact been conducted at different times, both more recently and with organizations other than just IBM. If these very different surveys detected the same pattern of differences between cultures, then per-haps we could be much more confident that the factors, in particular,

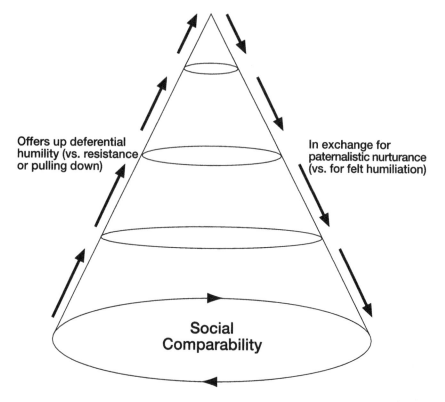

Figure 4.1 The working relationship between Hofstede's measures of high collectivism (vs. individualism) and power distance (vs. egalitarianism)

individualism–collectivism and power distance, are robust (see Gergen, 1973, for a fuller discussion on the importance of this question).

This is largely in fact what has happened. For example, a group of researchers calling themselves the Chinese Culture Connection (1987) surveyed students in twenty-two countries, using a values instrument generated indigenously, from and by Chinese sources. Both individualism–collectivism and power distance emerged in a subsequent country-level analysis (i.e., averaging items within countries). Trompenaars (1993) surveyed 15,000 managers (75 percent) and administrative staff (25 percent) of 30 multinational corporations in 50 countries, and found a negative correlation between individualism–collectivism and what he called 'Achievement–Ascription' (of status, which seems very similar to power distance). Hofstede himself has repeated his study, finding the same factor pattern repeated.

Most impressive of all perhaps, a recent meta-analysis of the predictors of conformity, across 133 studies in 17 countries (including Brazil, Fiji, Zimbabwe, Zaire, Ghana, and Lebanon), has sought to compare the predictive power of factors such as size of majority, nature of judgments being made, date of study, and how well the group members knew each other (Bond and Smith, 1996). Topping the list, in terms of predictive capacity, was Hofstede's country index of individualism–collectivism, with power distance (and Trompenaars's 'Achievement–Ascription') also being an important predictor.

Let us not also forget that aid assignments are very often about individuals, i.e., 'leaders' and 'followers', rather than entire cultures. What subtends at the cultural level does not necessarily apply at the level of actual *individual behavior.* At a cultural level, and due to the averaging process within cultures, behavioral *deference* to authority may go hand-in-glove with belief in (paternalistically) *nurturing* subordinates, whereas at the individual level, they are separated and even correlated *negatively,* because one person cannot defer and dominate at the same time! One study of 1,768 managers, in a single multinational company, found that only half of the variation between individuals could be accounted for by nationality itself (Griffeth *et al.,* cited in Giacalone and Beard, 1994). Individual differences are important.

Clearly therefore, we need to see some evidence that individualism–collectivism and power distance are meaningful and important *at the individual level.* That evidence can be found primarily in the work of Schwartz (1992, 1997). Along with various colleagues, and between 1988 and 1995, he surveyed values from 44,000 people in 47 countries from around the world. Schwartz focused on teachers, on the grounds that they play a key role in value socialization, conducting his analyses at the individual level within each country. He found, once again, dimensions corresponding closely to individualism–collectivism, as well as to power distance ('hierarchy versus egalitarianism').

Significantly for aid projects perhaps, Asian and Islamic cultural groups, in contradistinction to their Western counterparts, tended to believe in hierarchy rather than egalitarianism. Their particular definition of 'participation' (if they were to accept the notion at all, that is) would presumably be very different from Western definitions. This echoes our point made earlier, that definitions of participation may differ radically between potential donors and hosts, and compels us now to consider contingencies that are truly *cultural.*

Cultural contingencies

In Japan (Misumi and Peterson, 1985) and possibly in Australia too (Carr, 1996b), leaders and effective leaders seldom score a 'perfect' '9–9' on both task and relations (Blake and Mouton, 1978). Instead, they tend to be '7–7' leaders. Armed with the conceptual tools derived from Hofstede's (and others') analyses, however, we can see that Japan was/is relatively high on power distance while Australia was/is relatively low. In each case, it is culturally inappropriate to place oneself above others. That would mean becoming 'the nail that sticks out' (and 'gets pounded down') in Japan, and the 'tall poppy' (who gets 'chopped down') in Australia (see Chapter 9, this volume). In these two settings therefore, being a relatively good but not perfect leader may be the optimum to aim for (a more moderate version of this was discussed under trait theory). Hofstede himself (1980b) discusses how individual needs for power and achievement may not be too well received in more 'traditional' settings. Triandis *et al.* believe that individualists 'in collectivist cultures ... frequently find migration to individualist cultures necessary to escape the high demands that their ingroups make on them (1990, p. 1007).

What, then, of those indigenous leaders (or educated elites, as perhaps in Chapter 3), who attempt to apply their individualism as managers in the workplaces of developing countries? In Puerto Rico, for example, when management attempted to introduce a 'democratic' voice to every employee in a garment factory, the workers walked out (Marrow, 1964). They wanted a management that would make decisions and *manage*, thereby staying in business.

Similarly, perhaps, in another Puerto Rican garment factory there were no benefits to groups 'participating' in decisions; significantly better results were obtained when the group put forward leaders to discuss possible organizational changes (Juralewicz, 1974). The *meaning* of participation (and leader style) was/is *culturally* contingent. That is presumably why, in another developing country, India, managers who attempted to lead by 'democratic' and 'participative' methods were often rejected (Sinha, 1990). *Their* form of democratic participation was not what workers wanted.

Cross-cultural contingencies

Of course, the *real* difficulty in aid-project management, and TC, is that 'leadership' is not simply *culturally* or *contextually* contingent. It also has to be considered *across* donor and recipient cultures. As Hofstede has put it, 'North and West Europeans have trouble with the vertical society of nearly all Third World countries. They believe the first thing these Third World countries need is the elimination of their power inequalities' (1984, p. 395). In other words, the recipients of aid are often seen as needing some sort of health-giving 'dose of democracy'.

Is there support for Hofstede's conjecture in the aid literature? There, notions of 'democratization' and 'good governance' (and implied corruption and cultural inferiority) abound (Thomas, 1996). At a more local level, too, Thomas describes various 'conflicts of interest' between local NGOs, international NGOs, national governments, and local communities. These are often value-laden disputes about the 'social goals aimed at' (p. 102), that are not always solvable by democratic (rather than, say, negotiatory or power-brokering) means.

'Participation' in high resolution

Nowhere is the notion of social goals being set 'participatively' more clearly embodied, perhaps, than in the Western idea(l) of 'Management by Objectives', or MbO (Drucker, 1954). In the past, MbO is regarded as having *worked*, at least in terms of productivity in the West (for example, Level *et al.*, 1990; Rodgers and Hunter, 1991). It was intended to consist principally of setting, through one-to-one negotiation with one's supervisor, challenging but obtainable goals, and receiving continual feedback from that supervisor on one's progress, for example, through performance appraisal. Three hundred and sixty-degree feedback, as a form of performance management, involves feedback from alongside, above, and below in the organizational hierarchy. Compared with MbO, it is *extremely* participative!

Clearly, each idea or technique is potentially relevant to expatriate managers, insofar as their job is often to mentor themselves out of a job. On the face of it, either setting goals with a local counterpart, and/or getting them to evaluate their mentors ('upward' appraisal), would be very reasonable, commendable, and, above all, 'participative' models to follow.

However, the research reviewed above suggests almost the complete opposite (Hofstede, 1980b). The MbO ideal, for instance, would be heavily 'participative', in the Western sense, with value being placed on: (1) 'masculinity' (outcomes rather than processes); (2) job-related 'uncertainty *tolerance*' (taking risks in the job you are lucky enough to have, even though job security is often a big concern in developing countries, see Chapter 9, this volume); (3) individualism; and (4) low power distance (negotiate one-to-one with your own boss).

Even in a (moderately) power-distant French setting, Hofstede points out, *Direction par Objectifs* never sustained or indeed ever took hold in the first place. In many developing countries, as we have seen in Figure 4.1, the norm would clearly be to value power distance and collectivism. This would also render 360-degree feedback somewhat problematic. For example, a heightened sense of saving 'face' often makes any frank performance appraisals, let alone a full 360 degrees of them (!), unworkable (for example, Abdullah, 1992; Harriss, 1995; Seddon, 1985; see also motivational gravity in Chapter 9).

Thus, the question, 'Who wants to "participate" anyway?' would be far from fatuous. As Thomas (1996) informs us, there is an entire raft of possible forms of participation, many of them socio-culturally and/or socio-politically laden (for example, Freire, 1972; Sánchez, 1996). The valuable point that Hofstede's and colleagues' work has added to this literature is that management techniques, often built upon the Western concept of democracy, may appeal only to relatively egalitarian individualists, i.e., those who believe that everyone should 'do their own thing' and 'have their own say', as well as feeling relatively job secure and acquisitive.

A systems view

Overlaid on the cultural matrix (that the research seems to have described reasonably well) is an added dimension of cross-cultural *interactions*. Workplace relationships are not static – they change over time – and this inevitably heightens the importance of the degree of fit between the expectations of Western expatriates and the recipients of international aid (Klitgard, 1995). A vital question is how the host is likely to *react* to an expatriate's culturally influenced attitudes; how the expatriate will react to the recipient's reaction; and so on. It seems to us that a theory capable of addressing these *dynamics* must itself be dynamic. We now therefore offer a more dynamical, systemic view of the expatriate assignment.

In their discussion of expatriate assignments in the commercial sector, Giacalone and Beard describe two fundamentally different communication styles, 'instrumental' and 'affective' (1994, p. 624). These, of course, are broadly similar to the task and relations styles of leadership, but the authors suggest that the two styles may be culturally linked. It is conceivable, for instance, that an affective focus, in the long run, is more often adopted, at least on the surface, in some developing countries. As Harriss reports, for example, in an Indonesian context, 'It is unusual to apportion blame in cases of failure [for example] never be directly critical of individuals and never make the men . . . feel *malu* (ashamed, shy)' (1995, p. 120).

Giacalone and Beard suggest an interaction between these two styles of communication: 'When attitudinal differences between speakers exist, for example, it results in the instrumental speaker attempting attitude change in his or her receiver, whereas the affective speaker will simply change subjects' (1994, p. 624). In our view, this and its everyday workplace social analogues, could be the first step in an escalation of miscommunication and misattribution between expatriate and host. The theoretical and interactive process that we have in mind is pictured in Figure 4.2.

From Figure 4.2, we begin with a comparison of the degree of 'fit' between the two cultural backgrounds, particularly, perhaps, in terms of individualism–collectivism and power distance. In contrast to the host subordinate, the Western expatriate is likely to be relatively 'blunt' and direct – even confrontational – in their approach to evaluating work performance. They may, for instance, apportion blame individualistically, and do so as if the host subordinate were not expecting (or were entitled to) any kind of nurturance 'from above' (see Figure 4.1). In short, they are likely to want a more 'democratic' transaction.

From the host employee's point of view, however, the culturally appropriate tendency or reaction, as Giacalone and Beard suggest, is to attempt to avert social confrontation and social disharmony. Changing the subject, or avoiding it altogether, may be the result. To the expatriate, however, this starts to become irksome: the host employee will not confront their shortcomings or responsibilities! What seems to be needed, of course, is a healthy dose of even greater 'say' and 'participation' from this employee: they must be confronted, and made to confront their shortcomings. And so two vicious circles commence, with each side becoming increasingly defensive, but giving the impression of being noncooperative. Thus, MbO, or 'all-round' feedback, or its TC development-project equivalent attempt at 'democratization' and 'empowerment', fails.

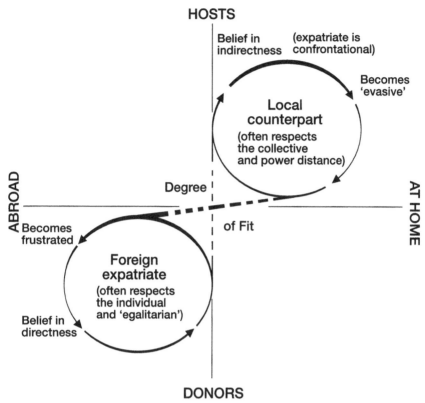

Figure 4.2 An escalating communication gap?
Source: adapted from Senge (1992).

Breaking the cycle?

Awareness techniques

It is theoretically possible to break such vicious and escalating cycles, and even to turn them into 'virtuous' ones (Senge, 1992). In the first instance, one might train expatriates (and host employees) in the appropriate ways of communicating, for example, through a series of simulated 'critical incidents' known as the 'cultural assimilator' (Brislin *et al.*, 1986). The assimilator was originally designed for people who are assigned at short notice, and was inductive, pragmatic, and context specific.

Another technique, suggested by Selmer (1996), entails asking about-to-be expatriates to estimate the work values of their about-to-be hosts, with these estimations being compared with actual ratings by the hosts themselves, in order to identify the points of least fit in Figure 4.2. In our view, and based on a more systemic perspective, Selmer's technique could be rendered more interactive, namely by involving the *host employees* more in the process, i.e., asking them to estimate the work values of the expatriates. As Zeira and Banai (1984) have pointed out, much expatriate selection has not taken account of host stakeholder expectations from an expatriate assignment, something that we have illustrated in our opening case study (see Chapter 1).

Building bridges

Transforming *vicious* circles into *virtuous* ones is a difficult task, but likely to be made easier by having good leadership research recognized by, and incorporated into, development studies. As we see it, there are reasonable grounds for believing that virtuous circles *can*, in principle, be socially engineered. As Kao and Ng Sek-Hong (1997) remind us, the so-called 'cultural divide' can in fact be bridged. For example, there is the reliable finding that the best leaders, the world over, somehow manage to combine task and relations concerns. To the extent that each of the partners in an aid project may tend to prefer one style over the other, there remains the possibility of cross-fertilization, particularly in view of Bales's findings. French and Raven's (1959) work certainly indicates, as we saw, that one can wield more than one form of power at once, or sequentially. The varieties of power (and leadership) are not mutually exclusive!

As these points suggest, moreover, it would be naïve in the extreme to imagine that collectivism and individualism are mutually exclusive. We are all collectivists at some times (for example, when cheering our national sports athletes), and individualists at others (for example, when *competing* as an athlete) and there is in fact a long-standing literature to support the fact that we can 'switch' modes, depending on social context (for example, from Sherif's early work on 'superordinate goals', see Chapter 9, and Tajfel [1978] to Turner [1991]), as well as the idea that one mode of behaving often frustrates the other. Scratch an individualist (or collectivist) and you may find a frustrated collectivist (or individualist) (Fromkin, 1972).

In our view, then, the terms 'individualist' or 'collectivist' should not be used to stereotype cultural groups. At best, they simply mean, at a group level or cultural level, that collectivists (or individualists) tend to spend relatively more time in 'group' (or individual) mode, or may be relatively difficult to shift *from* group (or individual) mode into an individualistic (or collectivistic) one (see, Turner, 1975). One should never make what Hofstede called the 'ecological fallacy', generalizing from cultural to individual levels as though individual differences do not abound and overlap considerably across different societies.

These commonalities (rather than differences) signal the possibility of escalating cooperation rather than escalating conflict. One of the most promising avenues for creating virtuous order seems to be found in the South American experience, of community social psychology and empowerment (Sánchez,

1996), much of which is based on the work of Paulo Freire (1972). In his review, Sánchez describes for us how, when an aid project began, its (outsider) 'leaders' perceived the local community and its representatives to be apathetic and lacking commitment to full 'participation'.

The solution in this case, and in others described in the review, was to clarify what was meant by participation itself. In that particular community setting, with fellow nationals (academics) as facilitators, a working definition of 'participation' was reached via group discussion and negotiation. In Hawaii, Klitgard (1995) describes how a project personnel's perceptions that 'recipients' were unmotivated were corrected after gaining a better understanding of, and implementing, culturally appropriate forms of participation, for example, working in and with groups rather than individuals. The same point may also (partially) hold in Aboriginal Australia (Hughes, 1988) and in Maori New Zealand (Thomas, 1994).

In the South American case, once some resolution of the issue of 'participation' had been achieved, each side began to value more and more the other's contribution to the project. There was an escalation of *cooperation*. The previously 'non-cooperative' recipients of aid began to volunteer contributions, for example, redecorating community-housing projects using traditional art forms. In terms of Figure 4.2, mutually defensive bluntness and evasiveness would have been replaced by mutually supportive and reinforcing consideration and initiation – by two *virtuous* circles.

5

INCREMENTAL IMPROVEMENT

Figure 5.1 The result of large leaps rather than incremental improvements
Source: Copyright © 1997, Matt Hodgson, illustrator.

> Among those who are aware of the term 'appropriate technol-
> ogy' there is an ever-increasing proportion who realise that
> Appropriate Technology is not just a matter of *things*. The early
> concentration of interest on *devices* such as windmills, on
> *materials* such as unfired clay brick, and on [technological] *pro-
> cesses* such as biogas generation is now being tempered by the
> realisation that Appropriate Technology is *socially* appropriate
> technology, and that therefore the human element is critically
> important.
>
> (Mansell, 1987, p. 1, emphases added)

The Orangi Pilot Project

Pearce (1996) has described the success of the Orangi Pilot Project in Karachi,
Pakistan. Orangi, a squatter district of Karachi, had been plagued by political
riots, murderous drugs mafiosi, and corrupt officials. However, it is now recognized as a shining example of how urban squalor can be overcome to produce
positive social change, with very little help from either local or national government, or from foreign aid. Akhter Hameed Akhan, a retired schoolmaster and
sociologist, who initiated the Orangi Project, found squatters living in homes
constructed from concrete blocks with limited water supply and no sewers. Residents of Orangi therefore emptied their bucket latrines into the narrow laneways

every four or five days. Typhoid, malaria, diarrhea, dysentery and scabies were rampant, and the children played in filth. Akhan decided that municipal authorities would never be able to provide appropriate sanitation for the poor of Orangi, and that foreign aid would not filter through to the poor. The only solution seemed to be *for the community to solve the problem itself.*

A brave starting point for this project was the rejection of existing technical standards as being too exacting and inappropriate to the social situation. A septic tank was installed between each toilet and sewer, so that only liquids reached the pipe, thus preventing them from becoming blocked and saving local streams from the worst pollution. New manhole covers were designed, smaller and simpler than conventional ones, and these could be produced in situ at one tenth of the price charged by outside contractors for conventional designs. There were no grants and no subsidized demonstration projects. Householders were simply encouraged to band together and install sewers *at their own expense.* UN consultants who visited the project in its early days found many faults with it: no targets, no surveys, no master plan. Orangi's sewer network was constructed in a piecemeal fashion. Individual lanes organized themselves to construct their own sewer, starting with those nearest the main streams and working up the hillsides.

Following the development of better sewers, the Orangi Pilot Project focused on improving housing, education, and health. Again the approach was to look at what already existed and how it could be developed, rather than imposing solutions from outside. The Orangi Project received some funding to pay project staff, but it also turned down offers of funding to build or set up 'a project'. This is because the ethos of the project is that people will look after something that they themselves have paid to develop. In addition, the project recognizes that subsidies can create conflicts within a community: if subsidies exist for some, then everybody will want one, and self-help will cease.

In Orangi, literacy rates have improved and are now twice the national average, sewage has been banished from the streets and health indices reflect dramatic improvements. For instance, in 1982, when the project began, infant mortality was 130 per 1,000 live births, but, by 1991, this had dropped to 37 per 1,000. Orangi also compares very favourably with Thikri, another squatter settlement outside Karachi, with no sewers and little organized preventive health care. Nine times as many couples use family planning in Orangi as Thikri, rates of illness are half that found in Thikri, and spending on doctors is only one-fifth of that in Thikri.

The Orangi success story started with sewage, and so it is interesting to note that the UN launched its own sewage scheme also in Orangi, in the early 1980s, based on a master plan and using conventional contractors. Pearce (1996) states that 'with a bloated bureaucracy, poor workmanship and inadequate maintenance it collapsed within five years, with just thirty-five lanes served'. By contrast, over 15 years the Orangi Pilot Project, working through principles of 'pay-your-own way self-help', with modest materials, making small-scale improvements, has installed 94,000 latrines, connected up to approximately 5,000 underground lane sewers and 400 secondary drains. Projects such as the Orangi Pilot Project give hope to the recipients of many failed development projects. They do this by emphasizing that communities which take charge of their own lives can determine their own future.

From sustainable change to incremental improvement

There are many reviews available of the factors in development projects which may contribute to sustainability (for example Bossert, 1990; Brinkerhoff and Goldsmith, 1992). But what is sustainability? Reviewing the sustainability of health projects, Laford and Seaman (1991/2) note that the Save the Children Fund has defined sustainability as 'the capacity of the health system to function effectively over time with a minimum of external input', while the United States Agency for International Development (USAID) define it as 'the ability of a health program to deliver services or sustain benefits after major technical, managerial and financial support from a donor has ceased'. These definitions suggest that sustainability is to do with creating something that will last. While it can be argued that 'sustainability' does not imply a fixed state (Bruntland, 1987), the word 'sustainable' means lasting, permanent, not perishable, not likely to give up.

The word 'change' means something else, it means making things different. So 'sustainable change' may be understood by some to mean making a difference and then keeping that difference. The supporting rationale seems to be that if enough resources are put into making a difference, then this difference should be rewarding enough for the various beneficiaries and stakeholders to ensure that the difference is continued, once the resources which created the difference are withdrawn. The problem with this is that when international aid workers encounter a situation of human impoverishment, they don't just want to make a small difference, they want to make a big difference! They want change on a massive scale.

Large leaps?

In development studies in general, and the study of aid in particular, several leading commentators have remarked that the *scale* of aid projects can be a bone of contention both with donors (for example, Dore, 1994) and sometimes with host governments (for example, Thomas, 1996). There is now a great deal of accumulated evidence, from educational counseling (for example, Downing and Harrison, 1990), cognitive psychology (Tversky and Kahneman, 1981), sociology (Cox, 1995), social psychology (Weick, 1984), management (Punnett, 1986), organizational theory (Braybrooke and Lindblom, 1963), and aid-project evaluations themselves (for example, Kaul, 1989; Psacharopoulos, 1995; Rondinelli, 1986) that it may often (although not always, for example, Dunphy and Stace, 1993; Gurstein and Klee, 1996) be more appropriate to pursue smaller-scale, incremental improvements (Posthuma, 1995).

This alternative approach, it has been argued, would provide a better match with traditions that are both valued (Gologor, 1977) and effective (Vittitow, 1983), and more generally with the idea that psychological changes often take longer than technological ones (Moghaddam, 1990). What sort of change the aid worker desires will no doubt be related to their personality. We suspect that many aid workers may be characterized as 'hero-innovator' types. Brunning *et al.* (1990) suggest that such people are heroic and charismatic characters who want to change the status quo and create a better world. They often expend a great deal of effort and energy on creating change. The problem is that when the hero-innovator leaves, unless there is another hero-innovator just around the corner, the energy to maintain the change they have created leaves with them. Possibly the hero-innovator's greatest failing is that he or she fails to create an environment which is capable of supporting and sustaining his or her innovations. The reason for this probably lies in the fact that the preparation and creation of such an environment is a long-term process, often much longer than the length of the development worker's contract. Often the development worker gives way to the temptation simply to get 'stuck in' and make some immediate difference, and the bigger the difference, the better.

A desire that things should change for the better and that there should be clear unambiguous *outcomes* from their efforts, can lead aid workers to overlook the *process* of change and the advantages of this change being anchored in local communities, through small and incremental improvements (MacLachlan, 1996a). We argue that people will be able to adapt to change more easily when the scale of change is small. While this is a simple message, we also realize that it may be contrary to the inclinations of many involved in international aid. Aid workers may seek to create 'large leaps' forward, from a status quo of what they perceive to be relative impoverishment; for example, they may prefer to establish a teacher-training college, rather than build one more small rural school. Also, the recipients of such aid, given the choice, may also opt for the larger-scale improvement, with its potential for delivering greater benefits. In each case these are understandable choices, both donors and recipients wishing to alleviate a situation of relative impoverishment to the greatest extent possible.

However, even though 'large leaps' may be desirable in some ways, we argue that this mentality is often the downfall of many development initiatives, and that a slower rate of change, through smaller incremental improvements, is a surer process for progress. We now review various sources of evidence, many of which are psychological, which together suggest that often the best solutions to development problems involve setting smaller, more modest, goals in the short term, and slowly incrementing these developments to achieve more substantial change in the long run. The general principle to be developed here is that *the more you change, the less you can sustain*. Even when this 'slower and smaller' approach to producing change is taken, change will always be a delicate and difficult process to manage effectively.

Field theory

Lewin's Field Theory (Lewin, 1952) describes human situations in terms of a balance, or imbalance, between opposing sets of forces. Stability results from a balance between forces for change (driving forces) and forces against change (restraining forces). However, an increase in driving forces, or a decrease in restraining forces will result in change, and such change proceeds through three stages. 'Unfreezing' is the first stage, and involves the dismantling of existing behavior patterns. 'Change', the second stage, is where people attempt to adapt to a new situation by adopting new behavior patterns. Thirdly, during 'refreezing', people integrate their newly acquired behaviors into their pre-existing repertoire of behaviors. An important feature of Field Theory is that it acknowledges that stability reflects a dynamic of interacting forces, not simply a failure of something to happen. Rural peasant communities in many developing countries are sustained through a way of life which may have changed little over hundreds of years. Yet this does not mean that such communities are stagnant. If instead they are seen as a 'successful' balancing act, between opposing forces, then the would-be 'developer' must approach 'community development' with a new set of questions.

Instead of asking 'Why have these people not "developed" more?', a better approach may be to ask 'Why has this situation continued to exist?' and 'What forces are involved in maintaining the balance here?'. These last two questions acknowledge the existence of an important dynamic prior to the 'development initiative'. It also acknowledges that interventions will necessarily produce change in more ways than one; however, the smaller the imbalance which is produced by change, the easier it should be to adapt to.

Stakeholders

If the status quo represents a balance between opposing forces, it also represents a balance between those people who control the forces. Such a 'balance of power' should not, of course, be taken as suggesting an egalitarian distribution of resources, instead 'balance' here simply implies equilibrium. An equilibrium can exist even when all the power is vested in one person (as in a dictatorship). A change in the status quo may well involve the empowerment of a subjugated group and this often implies the disempowerment of others. It therefore may make sense for those in power to resist change and avoid their own relative disempowerment. However, it is not only stakeholders who may resist change; even those who stand to gain from change may resist it.

Change, even a small change, may involve a step in the dark, giving up what is familiar and embracing the unknown. Various reasons for resisting change have been described by Bedeian (1980).

'Parochial self-interest' refers to the possible loss of respect, status, power, prestige, approval, or security which any individual who is a part of the change process may experience. This is because change is likely to affect extant social relationships and traditions. One's 'position' relative to others may change.

Another barrier to change may be misunderstanding the reason for change and lacking trust in those who are trying to create change. Such a situation can produce the feeling of personally being under threat, resulting in defensiveness and distorted perception of further attempts at communication. Another difficulty with the change process is that evaluations of its effectiveness may be contradictory, when presented from different perspectives. You may, for instance, be informed of the 'success' of a project when it has affected yourself in a very negative way. Different people, with different values, evaluate change in different ways. Clearly therefore, one of the constraints on the 'large leaps with bigger benefits' mentality, is the psychological cost of the scale of change.

Individuals differ in their ability to cope with change. It may be that those who are used to a slow pace of change can only effectively cope with changes which occur at a similar pace. In such cases, the lesser the magnitude of change, the easier it may be assimilated into people's lives. While these points have been raised by Bedeian in relation to organizational change in industrial settings, they can also be applied to change in development settings. Apart from the obvious organizational parallels to be found in international and NGO aid organizations, countries, regions, communities, and even families are all forms of human organization which may resist the process of change. Thus while we may think of 'stakeholders' as those people who explicitly have something to gain or lose (power, prestige, money, etc.) in the change process, it is clear that everybody who is likely to be affected by the change process is implicitly a 'psychological stakeholder' because the process may change how they relate to their social, economic, or physical environment. There is a risk involved in the change process even for those who may be quite unaware of the changes. This is because individuals' and groups' position relative to sources of power and influence may also change. Thus those unaware of the change process may find themselves disempowered or empowered by it, without being able to determine their own role within it.

Change as risk

To engage in a process of change necessarily creates a situation of risk. A willingness to accept risk varies from one person to another. Sometimes organizational psychologists are contracted to identify individuals who feel either very comfortable in a high-risk environment (for example, bomb disposal) or very comfortable in a low-risk environment (for example, civil servant). The point is that different people are comfortable with different levels of risk. There is some evidence that tolerance for risk-taking also varies across cultures, with some cultures being described as 'high uncertainty avoidance' (i.e. risk aversive) cultures, and others being described as 'low uncertainty avoidance' cultures (Hofstede, 1980,

1991; see also Chapter 4). Operating in the region of the unknown is more acceptable in some cultures (and subcultures) than in others. Change involves risk because it involves entering into the unknown.

Entrepreneurs are often seen in the context of a capitalist society as individuals who are willing to take risks by taking business initiatives; for instance, in trying to open up new markets for a product, or trying to develop a new product for existing markets. However, Mansell (1987) points out that there are many types of entrepreneur, and he includes in these, so-called 'community workers', 'change agents', and 'extension workers'. They too are engaged in risk-taking to change the status quo: they are 'social entrepreneurs'. Mansell gives an example of a teacher in a rural school recognizing the potential of a nearby stream to produce energy in the form of hyrdo-electric power, thus creating a means to better lighting, heating of water, and use of power tools. The teacher manages to acquire financial support and technical assistance from other villagers and some support from outside of the village. He organizes labour to put the hydro-electric scheme in place, and an organization to maintain, operate and administer the scheme.

Mansell (1987) suggests that some of the risks inherent in this initiative include: (1) borrowing money; (2) building the hydro scheme; (3) introducing a fee-for-service (probably on credit) scheme, when perhaps no such scheme has been used in the community before; (4) introducing the need for new skills within the community; (5) disturbing relationships with other people in the community who make a living out of alternative ways of creating light, heat, or undertaking strenuous work, and so on. Should the scheme fail, the teacher may be left with hefty debts, a loss of social status and a lack of self-confidence. Furthermore, impoverished communities, almost by definition, have very little capacity to absorb such failure. It may therefore be in everybody's interest not to engage in high-risk entrepreneurship. One way to reduce the risks involved in change is, as we have stated already, to make the degree of change small, so that successful change can be more easily accommodated and unsuccessful attempts at change can be more easily absorbed. So, while impoverished communities may desire improvements and therefore changes in their living conditions, it may be better for them to engage in a series of small incremental changes, building up to a greater long-term effect.

Small wins

Weick (1984) has argued that people often define social problems in ways which overwhelm their ability to do anything about them. For instance, take the problem of hunger in the USA. To reduce hunger more food should be grown, which requires greater use of energy for farm equipment, transportation, and fertilizers, which will add to the price of inputs and therefore increase the price of food, which will put it out of the price range of the needy. Another example given by Weick is solving the problem of rising crime. One response to this is to increase the number of law-enforcement personnel. But to do this means reducing the funds available to other areas such as schools, welfare services, and job training. This in turn leads to more poverty and less opportunity, resulting in more crime, more prostitution, and more drug abuse.

Scaling-up of problems can paralyze efforts to overcome them. This time let us take an example from international aid. Why are the majority of people in sub-Saharan countries living in poverty? One construction of this problem is that developing countries are being encouraged, through structural-adjustment programs, to compete in a global market. These countries may have a market advantage through the supply of cheap labour. Because labour wages are low, people are trying to get by on a day-to-day basis and are not able to save money, or invest their money in health, education, or other services. Furthermore, competing in a global market means that you are vulnerable to changes in the market, anywhere in the world. Attempts to improve the standard of living by increasing the wages in one developing country may give another developing country an economic advantage, thus taking the trade away from the country which increased the wages of its workers.

Scaling-up the gravity of social problems also increases the importance of the issue at hand, and so the magnitude of the demands to cope with the situation. What then happens to the perceived capability to cope with these demands? Based on a substantial amount of psychological research, Weick has argued that when perceived demands exceed perceived capacity to cope, then stress, in the form of frustration, arousal, and helplessness, ensues and these reactions result in poor performance and thus poor problem-solving.

One of the classic findings from experimental psychology is that an optimum level of arousal exists for performance. At too low a level, people are bored and do not fully engage in their work. At too high a level of arousal, performance also deteriorates. This can occur in several ways. For instance, people revert to old and familiar patterns of behavior on a task which requires novel and innovative approaches for its solution. The point here is that because of the high level of stress induced in trying to manage a highly arousing task, people fall back on what they know, even if it doesn't work. What is crucial therefore is that the task people undertake is challenging enough to prevent boredom, but not so demanding that their ability to cope effectively is diminished. Instead of seeking big changes, it may therefore be wiser to seek a series of small changes. This logic has been applied to technological innovations in the USA:

> A technological innovation is a big step forward in the useful arts. Small steps forward are not given this designation; they are just 'minor improvements' in technology. But a succession of many minor improvements adds up to a big advance in technology . . . It is understandable . . . that we eulogize the great inventor, while overlooking the small improvers. Looking backward, however . . . It may well be true that the sum total of all minor improvements, each too small to be called

an invention, has contributed to the increase in productivity more than the great inventions have.

(Machlup, 1962, p. 164)

The same argument may be made for technological, or indeed any other form of improvement, in less industrialized countries. International aid has too often focused on the size of the (perceived) gap to be bridged between 'developed' and 'developing' countries, and in doing so has overlooked the value of small wins. Technological cathedrals have been erected in social deserts, their sustainability being undermined by the gap which they open up between local resources and 'Western'-inspired aspirations. Rather than bridging the gap, *technological leaps are the gap*. They often constitute the problem and not the solution. Many technological innovations in developing countries reach beyond the carrying capacity of the local economic, social and educational infrastructure. The inability subsequently to 'service' such innovations may demoralize local communities, their value having been invested in their ability to sustain such 'innovations'.

The notion of small wins has much in common with incremental improvements. Weick characterizes a small win as 'a concrete, complete, implemented outcome of moderate importance. By itself, one small win may seem unimportant. A series of wins at small but significant tasks, however, reveals a pattern that may attract allies, deter opponents, and lower resistance to subsequent proposals. Small wins are controllable opportunities that produce visible results ... Once a small win has been accomplished, it forces a set [an attitude] in motion that favors another small win' (1984, p. 43). This is a philosophy that lends itself to the resource-depleted environment, which, almost by definition, characterizes the situation of developing countries. Here the message is not to bridge the gap, but instead gradually to build. If we accept the idea of aiming for small wins, we then need to establish just how small can an increment be, and still be meaningful? Some guidance in this regard comes, somewhat surprisingly, from the field of study called psychophysics.

Making a just noticeable difference

Psychophysics is concerned with the measurement of physical aspects of psychological experience. In 1834, E. H. Weber found that *discrimination* of stimuli is *relative* (Schiffman, 1982). That is, our ability to discern a 'Just Noticeable Difference' (JND) or increase in, say, candle numbers is proportional to the magnitude of stimulation at the time. Adding 1 candle to 60 may be just noticeable, but adding one to 120 is probably not. By analogy, then, what would count as a JND, from a relatively and materially deprived host perspective, may be a whole lot less than what would count as a JND for a comparatively wealthy government minister, or donor. Their *threshold* for a JND would be much higher – possibly partly prompting them to call for a large leap.

Weber's finding, which generalizes across all the senses was developed into Fechner's Law, which is visually depicted in Figure 5.2. According to Fechner, a given JND corresponds to a proportional increase in the stimulus. With background intensity at a low, the increment for a JND is small, but with background intensity high, the stimulus increment necessary for a JND to occur will be larger. Fechner's Law is also described algebraically as follows:

$$JND = {}_\Delta S = C \frac{\Delta R}{R}$$

where ${}_\Delta S$ stands for a change in sensation, R stands for *Reiz* (the German word for stimulus) and C is a constant which varies from one sensory modality to another. We believe that the relationship described by Fechner can be usefully reinterpreted in terms of development, as

$$JND = C \frac{\Delta DV}{DV}$$

where JND stands for Just Noticeable Development, DV stands for a Domain Valued (that is, an area of activity, or domain, which is held to be important, or valued, by the recipients of the development initiative), and where C stands for

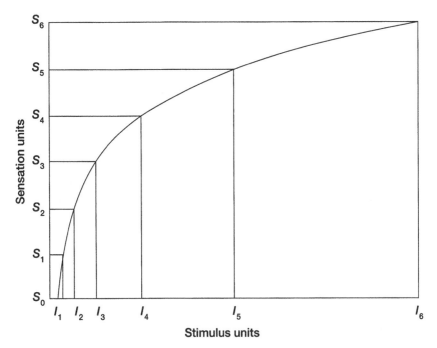

Figure 5.2 Fechner's Law
Source: adapted from Schiffman (1982, p. 17).

93

the broader Context in which the intervention is taking place. Clearly, different contexts will require different levels, and indeed types, of input to create Just Noticeable Developments. So, for example, a small investment of resources in a drought situation may make a large difference, while the same size of investment in improving the hygiene facilities of a large town may make very little, or no, difference at all.

In relief situations, where the extent of human misery is extreme and sustainability is less important than immediate survival, it will be important to create a big difference in people's lives. However, in development situations, where the long-term maintenance of positive change may be seen to be more important, we believe that making a just noticeable development may be the most effective intervention in the long run, assuming that it is followed on by other just noticeable developments. A just noticeable difference is that level of increased stimulation (or input) where people only just recognize that something has changed. Thus, in the case of aiming to make incremental improvements, 'developers' should seek to keep the fraction $_\Delta DV/DV$ as small as possible, while retaining enough of a change in the valued domain that people recognize something positive is happening.

Once again, to know what would constitute a just noticeable development for a community deprived of a clean water supply, we need to go to the community and learn from them. This does not mean that we always have to go along with, perhaps the preferred 'gift', of, for example, giving new shoes, *rather than* fulfilling the developer's aim of, for example, restoring sight. Indeed, there is a moral imperative on all of us engaged in trying to improve the well-being of others, to explain alternatives and advocate actions which we believe to be most beneficial, but there is also a moral imperative *to listen to, and learn from*, the informed judgments of others. If the additional price of restoring failing vision must be a new pair of shoes, then so be it! Incrementing just noticeable developments focuses on the *process* of development rather than its outcome. It will be an anxiety-provoking venture for those who prefer the comfort of demonstrating significant gains by the end of three- or five-year development projects. However, those whom aid workers seek to benefit are not so restricted. They seek only a better life, not necessarily a better life by the expiry of the current contract (!), which has been successfully tendered for by the current development agency.

The psychotherapy analogy

Within the realm of psychotherapy, the problem of preventing relapse, given that treatment has been successful, is perennial. In this context relapse refers to returning to an undesirable psychological state which is distressing to the person experiencing it and/or to others. Ager (1988, 1991) has drawn a comparison between the problem of preventing clinical relapse and the problem of sustaining change in development projects, where the site of change may be a community, or a particular service, rather than an individual patient. Adopting a behavioral perspective, Ager stresses the importance of establishing an environment which is capable of maintaining the new desired behavior. This 'antecedent environment' must be capable of providing reinforcement of (or reward for) the desired behaviors.

In order for the behaviors to be maintained (for instance, the continued use of a new stand-pipe for water in a village located close to a fresh-water lake), they must be reinforced *consistently* (each time people go to use the tap it works), *reliably* (people can depend on the tap working) and that they themselves have *control* over the tap (as opposed to other people operating it and therefore being able to determine who has access to it, how much they may take and so on). Ager also described important 'consequent conditions'. That is, the conditions which come about as a result of using, in this case, the stand-pipe for water. Here the reinforcers should be *natural* (instead of contrived, as is the case when tokens are given for good behavior and these tokens may be traded at a later date for some valued good or activity), *functional* (practically useful) and *appropriate* to people's needs. In the case of water being made freely available, which can be seen to be not only more hygienic than lake water but also more convenient, the recommended 'consequent conditions' for the maintenance of the desired behavior of tap use are easily achieved.

This scheme may therefore seem somewhat simplistic in the case of tap water, however, if you consider the case of maintaining a behavior change, such as vaccinating new-born children, it is much more complex. Here the challenge is to create an environment which will encourage and maintain vaccination behavior, using reinforcers which will ideally have the characteristics of consistency, reliability, and control, and be natural, functional, and appropriate. Another important element in maintaining an acquired behavior, is that its 'response cost', should be as low as possible. The desired behavior, if it is to be maintained, should 'cost' as little as possible. Here, cost may refer to any difficulty or loss an individual may incur through performing a certain behavior. Costs may be calculated in terms of time, money, energy, status, and just about anything else which people value. In practical terms, 'response cost' is therefore related to the economist's 'opportunity cost' and the idea that if you invest resources in one thing, then you lose the opportunity of investing those same resources in something else. These ideas are increasingly seen as having currency in the development literature through so called 'time–energy' studies. As we have already noted, engaging in and maintaining a new behavior is not simply a matter of 'adding something on' to your life; rather, there are often forces opposing behavior changes, and the cost of unfreezing these may be unacceptably high.

The behavioral approach has the potential of taking into account the broader environmental context in which behavior change must take root. It also has another advantage in that many of its therapeutic techniques are based on the principle of slowly building, one step at a time, in the desired direction, rather than making great leaps forward, as may be the desire of the troubled client or patient. The rationale is that by slowly building up a skill, the person acquiring it has the opportunity to integrate it into their existing repertoire of behaviors. Thus Ager (1988) has also argued for the potency of minimal therapeutic interventions as a means to creating incremental change. The technique of Forward Chaining, commonly used to shape new behaviors in people with learning disabilities, is a good example of making improvements incrementally. Here, a behavior, such as drinking soup from a bowl by using a spoon, may be acquired first by teaching someone to pick up a spoon, then placing the spoon in a bowl of soup, then lifting the spoon out of the bowl, then bringing it toward one's mouth, opening the mouth and so on. Each of these steps, in the case of someone with a severe learning disability, may take many days or weeks to master, and sometimes, sadly, they cannot be mastered at all. The point of this example is not to suggest that the methods of development practice should treat the recipients of development projects as if they were in some way 'simple'. The point is not about the recipients, it is about those trying to change behavior: the process of change may need to be slow in order to allow people to integrate what they are being asked to do, with what they already know how to do. The 'great leap forward' aspirations of international aid workers may well seem appealing to relatively deprived communities in developing countries. However, no matter how valued the ultimate goal is, small steps will create a surer path to achieving it.

This chapter opened with a quote about the importance of the human factor in technology transfer. Mansell (1995) argues that technology transfer is not simply about making new devices available, it is also about transferring ways of doing things. Sometimes such transfers have an implicit assumption that the recipients do not know how to do the things which the devices are designed to do. However, societies long predating modern science had their own technologies, which were perhaps non-scientific rather than un-scientific. We often think of technology as being scientific; however, this is not necessarily so. Inasmuch as technology is about finding means to perform certain functions, all societies – scientific or otherwise – have developed their own technologies. 'Technology transfer' may therefore often be experienced, not as new technology, but as technology replacement or updating. If sensitively introduced, it may still be very welcome.

Whose standards?

Mansell (1978), an engineer, describes the use of aerial ropeways in Papua New Guinea to carry goods to remote villages which are inaccessible by road. These ropeways are also used by people. A ropeway designed to carry a heavy commercial load is probably also capable of carrying a person. However, in the West there is a great difference between the design of ropeways for carrying people and that for carrying goods, with the former having much more stringent, and much more expensive, design criteria. So, for instance, the importation of Swiss standards of performance may prevent the development of new ropeways in Papua New Guinea, because the cost of conforming to much higher design specifications makes them unsustainable in the local context. Mansell states that while the optimum margin of safety against the breakage of a rope may be the same every-where, different contexts require different degrees of comfort and operational reliability:

> Failure that leaves the ropeway immobilized for some time may cause intolerable suffering to pampered European tourists and may justify a relatively expensive design for passenger-carrying equipment. In a developing country it is very likely that the standard set for goods-carrying equipment is acceptable to people whose alternative means of transport is to carry their goods on their backs on steep mountain paths.
>
> (1978, p. 278)

Mansell does not deny the very real moral dilemmas that exist in trying to balance economic viability and acceptable levels of safety. Rather his point is that the internationalization of standards, by which development projects may be constrained, can actually prohibit the ability of such projects to affect as many people as they would like to. If only one out of a possible ten villages has the benefit of a 'world-class' ropeway, how many people may fall from steep moun-tain paths on their way to the other nine villages? Smaller, more modest improvements, taking the local (rather than international) context into account, may produce greater benefit, both economically and in terms of providing a safer mode of transport for a greater number of people. This same point was also exemplified in the Orangi Pilot Project's rejection of existing technical standards as being too exacting and inappropriate to the social situation in Orangi.

Cultural resistance

There is another, more bothersome, aspect of the aforementioned behavior ther-apy and development aid analogy. Often times the desire for somebody with a learning disability to acquire a new behavior comes not from him or herself, but from someone else taking a paternalistic role. Now in the case of helping a learn-ing disabled person to function more effectively, some such interventions can be justified. However, in the case of development aid, it is highly questionable. Clearly the targets for development should not be determined paternalistically by those not engaged in the problem (*if there is a problem*). Instead, they ought to be born of the people's desire to improve the circumstances in which they live. This may seem naïve, because those who are deprived may not be fully aware of the degree of their relative deprivation, the means available to ameliorate it, or the alternatives for which they could strive. But this is not to endorse a paternal-istic approach which ultimately serves the needs of aid agencies, rather than those they claim to be assisting. The necessary *tension* which exists between paternal-istic and self-growth philosophies of aid may express itself in development pro-jects through apparent 'resistance' on the part of the aid recipient.

Resistance to development initiatives – even when they may be seen to be of obvious benefit to the people resisting them – may be a form of negotiation, where the recipient is struggling to retain a sense of self-respect and inner control (see Chapter 11). Verhelst ([1987] 1990) sees many theories of development as 'catching-up theories', with an implicit assumption that societies which do not catch up are destined for extinction. Verhelst also recognizes a form of resistance from communities in developing countries toward the process of 'development'. One function of such resistance is that it may slow down the process of change, thus making it more manageable. We are not suggesting that people consciously choose to slow things down, but that actions (or lack of actions) which impede the progress of development projects may actually have beneficial effects for the recipients of aid. Through a plethora of inefficiencies, distractions, and demotivations these recipients may force the pace of change to progress incrementally, in smaller and more easily assimilated steps, than aid workers may feel comfortable with. Again, it is important to emphasize that it is not that the recipients of aid don't want a much better material existence, rather that they need to retain a sense of socio-cultural identity and cohesion determines that a slower pace of change will be necessary. As noted earlier, the social mechanisms for change take longer than the technological mechanisms. Consequently, we would predict that where people have made large leaps in technological development, in time they are more likely to have left behind their cultural identity and be experiencing a greater degree of cultural anomie, than those who have changed more slowly.

Developing a similar theme, Verhelst states that 'the origins of this kind of resistance may well lie in the cultural uniqueness of each of the populations in question and in their need to safeguard their identity' (p. 16). We believe that many of the phenomena we have studied (for instance, the 'Pay Me!' reaction [see Chapter 11], motivational gravity, and 'double demotivation' [see Chapter 9]) may also reflect, in part, cultural resistance. In psychoanalytic clinical practice, *resistance* refers to the situation where the therapist makes an interpretation of their patient's behavior, and the patient denies the validity of the interpretation: 'No I don't think you're right about that.' Rather than the therapist doubting him or herself, he or she may take the denial as proof that they are right: 'You are saying that because what I have said is painful for you to come to terms with' (put another way, 'I'm right, whatever you say!'). This paternalistic and arrogant stance surely resonates with some of the attitudes that have been identified as causing aid projects to founder.

Valuing change

What is valued in one culture may not be valued to the same extent in another culture. For instance, in some cultures the distribution of resources in response to need is seen to be the most powerful factor influencing decision-making, while in other cultures, the distribution of resources in proportion to individual effort may be the most popular choice (Berman *et al.*, 1985; Chapter 6, this volume). This example simply serves to illustrate that since value hierarchies differ across cultures, beliefs about what should change, and how it should change, may also be dependent on culture. We have already argued that many development projects reflect the values of aid workers rather than those whom they seek to assist. Notions of appropriate and desirable change will differ, not only from one culture to another, but also from one perspective to another: for instance, the same sort of 'aid' may be experienced quite differently from the giver's and receiver's perspective (see Chapter 9).

To know whether an incremental change will truly constitute an incremental improvement, we must first learn the values of the people we seek to serve. It may, for instance, be distressing for us to learn that an elderly man values having a new pair of shoes more than he values having his failing sight restored. If we are to be truly led by the values of the communities which aid projects seek to serve, then we must understand the social context and meanings which generate these values. Perhaps for the elderly man referred to above, the social status of new shoes is more important than the ability to see his own impoverishment, or to see people's awareness of his impoverishment. If so, he will not be alone. Change will only be valued if it allows individuals to be 'somebody', as opposed to being the 'recipients' of aid (see Chapter 11). Community, as has been convincingly argued by Freire (1970), can be an important vehicle of social identity and empowerment.

Community

The origin of the word 'community' comes from the idea of sharing a wall. Such a wall, being a barrier, functioned to keep some people out, but at the same time defined what was common ground to those protected by the wall. In ancient times, this sharing of space referred not just to physical space, but also to the psychological space created by the sense of enclosure. People shared not only the physical space enclosed by the wall, but also the psychological responsibility of cohabiting within that space. Thus a community is as much an organ of exclusion as it is one of inclusion (Tajfel, 1978). Many developing countries, especially in rural areas, retain a strong sense of community and, while this ought to be harnessed as a vehicle for change, it is often ignored or even problematized.

Community values (like cultural values) are often seen as a source of resistance to change. However, given the above account of what a community is, it is hardly surprising that change should be resisted, when it is sanctioned by strangers outside of the community. Change must come from within. There is an old joke that goes something like this: 'How many psychologists does it take to change a light bulb? Only one, but the bulb has got to really want to change!' A similar principle may be applied to development projects: how much capital, hardware, and human resources have to be poured into a community to produce a change? Answer: very little, but the community has to want to change.

PSYCHOLOGY OF AID

The psychology of community rehabilitation

It is also important to recognize, however, that many communities in developing countries are not doing well. We cannot afford to be precious about the importance of retaining community and cultural values if the way of life they promote is not serving people well. In a changing world it is inevitable that some practices and values will have to adapt to new demands. The folklore of aid is embellished by many outlandish tales of projects employing a nutcracker to peel an orange, so to speak. Inappropriate technology is used to overcome 'problems' which the recipients of the aid never even considered to be a problem. When real problems do exist, development workers, rather than seeking to solve the problem, may try to transform it into something for which they already have a solution. Much development work proceeds through what we have called the *transplant mentality* (MacLachlan, 1996a). The analogy is that the 'developed world' is donating a vital organ (money, equipment, or personnel) to the developing world. The right sort of donation, at the right time and in the right place, will rejuvenate the 'frail and fading' recipient.

In a real organ transplant, however, the procedure is complicated by the fact that the recipient's body recognizes the donated organ as being *foreign* tissue, not part of itself, and consequently fights against the new organ, attempting to 'reject' it! If the organ is accepted, the diminished effectiveness of the immune system (achieved through high doses of medication designed to immobilize the body's defenses), leaves the patient open to other dangers. An 'accepted' organ may also demand too much of the patient's body, lacking the necessary resources to carry it.

The efforts of Western biomedicine to have an impact on the health problems of many developing countries has often been akin to this transplant mentality. The biomedical approach to illness is often seen as foreign and, at least in part, may be rejected by many. The infrastructure of Western biomedicine, comprising of new hospitals and technologies, often consume a disproportionate amount of a country's health budget. Furthermore, when international aid is withdrawn, the indigenous system is not capable of providing the level of resources necessary to sustain the hospitals and technologies. These leaps in development, by undermining and ignoring extant community resources (for instance, a network of traditional healers), often actually worsen an already poor situation, by making communities dependent on outsiders' continuing to supply 'aid' to support their donated 'organs' of development. The withdrawal of aid may leave communities in a worse situation than they were 'pre-development', as local capabilities and resources are diminished through competition with the service that was being developed. This can in turn result in even greater needs, and greater dependence on outsiders, who in turn concoct more and more ambitious projects in order to redress a situation of escalating impoverishment. The patient, the aided community, can in this way, free fall into the development process. Development aid, when administered badly, may not simply fail to make an improvement, it can, and has, made its recipients worse off.

100

An alternative approach to the process of change, which we have tried to advocate in this chapter, is more consistent with a *rehabilitation* analogy, describing the need to regenerate growth in a damaged organ. We do not want our rehabilitation analogy to be taken as suggesting that the community is 'sick'. Instead, we wish to recognize that in some developing countries the way in which people live, and the environment in which they live, constitutes human deprivation. We wish to emphasize that often these communities themselves have the potential to remedy their own problems and, possibly, without donor assistance.

This analogy suggests that what already exists can be improved upon, rather than 'cut out'. For instance, a community may endorse ways of working which are less adaptive now than they have been in the past. Rehabilitation of community values and practices, in a slow incremental fashion, can save us from sacrificing the integrity of the community. The focus in this, more psychological, approach, is on analyzing what already exists (a particular behavioral repertoire perhaps) and improving its functioning. Building from strengths and incorporating community virtues must be key ingredients in the incremental improvement process.

Finally, there is an aspect of the incremental-improvement process which we need to face up to, because it is potentially uncomfortable for the development worker. If the appropriate targets for change are to be identified by members of the community rather than aid workers, then aid workers may not know exactly what it is they are going to be 'developing'. If aid workers are able to tolerate this sort of disempowerment and still facilitate a process of change, then they may well find that subsequent increments are not the sort of things that they had expected to be 'developing'. The aid worker may be taken off in directions never anticipated, and ones which may not clearly lead anywhere (see Chapter 3). Indeed, sometimes they may lead nowhere, and sometimes they may produce solutions to problems which the aid worker did not even know existed. Ultimately, the aid worker, and his or her head office, are going to need to develop faith in *pursuing the right process of change*, rather than seeking to achieve some predetermined outcome. It is salutary to note that UN consultants who visited the Orangi Pilot Project in its early days, were dismayed to find no targets, no surveys and no master plan!

6

EXPATRIATE WORK MOTIVATION

At least one-fifth of all overseas development assistance is spent on technical cooperation (TC), much of it on expert, i.e., 'expatriate', aid (Cassen, 1994). In sub-Saharan Africa, the proportion reaches one-quarter, and it costs on average US$300,000 per annum to fund a single expatriate (Dore, 1994). With this kind of money being invested, donors would want to be absolutely sure that, as a form of aid, TC is proving relatively effective. That is especially so since some aid agencies pay their expatriate employees a local salary. Yet evaluations tend to be short term in focus as well as in-house, and often left to the expatriates themselves. As a result, as Cassen observes, they may not get done, or run the risk of containing success rates that are artificially (and self-servingly) inflated.

Although there *have* been some clear successes, notably in engineering and population-policy projects in Asia, the record overall is mixed regarding the institutional and human-resource development that comprises the bulk of TC. This is particularly so in sub-Saharan Africa (Cassen, 1994, p. 146), where most money is also spent. As we saw in the Tanzanian case study (see Chapter 1), an expatriate aid project there also failed badly, a failure that was helped along considerably by the *very amount* of money spent on the expatriate's salary.

In neighboring Malawi, as indeed elsewhere in the developing world, it is not unusual for expatriates to draw ten or twenty times the local pay (for example, Fox, 1994), often with an additional local salary. Such differentials will inevitably strain the relationship between expatriate and local counterpart. Indicative of that, perhaps, expatriate–host relations have long been the subject of a rich and clandestine folklore (Gow, 1991), which can easily become highly distortive in cross-cultural contexts (Bartlett, 1995): 'The evaluation literature frequently singles out the expert, and the expert–counterpart relationship, as a key element of TC in need of strengthening' (Cassen, 1994, p. 169; see also, Gergen and Gergen, 1971; Sahara, 1991; World Bank News, 1996).

Often, we believe, the blame for aid project failures is laid, anecdotally or otherwise, at the door of *host work motivation*. In this chapter, however, we 'turn the tables', so to speak, presenting new evidence evaluating the consequences of expatriate–host salary differentials on *expatriate* work motivation. The fundamental attribution error, for instance (see Chapter 2), cautions us to avoid blaming the victim, and suggests that we should sometimes look more closely to ourselves. According to Cassen, among the complaints anecdotally heard against aid expatriates are: failing to mentor, or ever to select, a local counterpart; upstaging the local counterpart; blocking the counterpart's career progress by staying too long; failing to report on the assignment (the latter two may be related); and selecting a counterpart too late in the life of a project. Our own emerging data are suggesting that the differential between aid salaries and local salaries may be a key factor in each of these complaints.

If (as we will argue) higher wages do not necessarily mean higher work motivation, this raises the question of whether scarce resources could be allocated more effectively. One of the possibilities identified as promising in the aid literature is increased recruitment from the developing countries themselves, with salaries determined by the local not the international market (Allen, 1987). We therefore end this chapter by considering the growing number of expatriates originating from non-Western, so-called 'industrially developing countries' (Ong, 1991), highlighting some of the alternative pitfalls that they may encounter, as well as how (possibly) to manage them more effectively.

Why do people work?

The answer to this basic question is by no means as simple as it may seem. Common sense seems to dictate that people work mainly for money, and that there ought to be a relatively straightforward and positive relationship between pay and performance (Beacham, 1979). This is partly perhaps why managers and administrators often assume that financial motives principally govern workplace attendance and behavior, a widespread view of work motivation that is known in the management literature as 'Theory X' (McGregor, 1960). This 'carrot-and-stick' view of how to motivate workers reportedly permeates organizations and managers operating in developing countries (Blunt and Jones, 1992; Kiggundu, 1990), including perhaps some expatriates working for the international aid agencies. People often seize upon the most obvious explanations for behavior (for example, money, backed perhaps by the attribution that people are not naturally disposed to work) and prejudicially discount other possibilities (Kelley, 1972).

Addressing this issue empirically however, researchers in many countries have asked people whether and why they would continue to work in the absence of financial necessity (for example, Carr, MacLachlan, Kachedwa, and Kanyangale 1997; Chan et al., 1992; Harpaz, 1989; Jones et al., 1995; Meaning of Work (MOW) International Research Team, 1987; Morse and Weiss, 1955; Warr, 1982). The world over, it seems, most people(s) would want to continue to work. They tend to have non-financial needs that work also satisfies or must satisfy. For example, it frequently entails earning social and self-respect, *as well as* financial security. In poorer countries, job insecurity may focus indigenous people on material concerns (Blunt and Jones, 1992), but a search for social identity may continue to exist in parallel with that concern (Carr *et al.*, 1997). The general problem with Theory X, therefore, is that it does not consider sufficiently *the complex plurality of the human factor.*

In particular, it does not allow for the fact that money is often necessary, but not sufficient, for self-fulfillment (Argyle, 1989; Herzberg, 1966; Mayo, 1949), a managerial view that McGregor (1960) termed 'Theory Y'. In particular, employees from the more industrialized economies (like expatriates) have come to regard reasonable wages as a basic entitlement (for example, Hartman *et al.*, 1997), leaving them relatively free to dwell on certain 'higher-order' concerns, such as the extent to which their job gives meaning and purpose to life. This search for a meaning in work, we submit, is liable to be especially true among expatriates working in aid.

MacLachlan (1993a) has detailed both the tremendous differences in lifestyle between 'aidies' (internationally salaried aid workers) and the people they are meant to serve, and the resultant motive to distance oneself, psychologically at least, from those sharp (and proximal) privileges. Considerations of social justice may thus be impacting on aid-expatriate motivation and performance. Logically perhaps, that would be especially so among those expatriates who are most clearly motivated by altruism. In an extensive study of Canadian aid expatriates, motives ranged from 'wanting to help others' to 'going because the money is great' (Kealey 1989, p. 397). Yet it was the *latter* that foretold the most ineffective skills transfer. We therefore begin our analysis with that, presumably less committed, category of aid expatriate.

From self-perception to equity restoration

According to Bem (1972), people who are not particularly committed to an action will often use their own behavior to infer their own motives. Having nonchalantly bought and tried a new brand of toothpaste, for instance, we might infer that we like it *because* we bought and tried it. Essentially, Bem argued that when we are low on initial commitment and attempt to understand our own behavior, we resemble observers of others' behavior rather than actors in the behavior itself (see Chapter 2). As Littlewood (1985) observes, expatriates do sometimes lack commitment to the people with whom they will work. In keeping with his predictions, Bem (1967) found that compliance with a boring, low-commitment task under high-incentive conditions resulted in very money-centered attributions.

By an analogous process, perhaps, expatriates who begin their assignments with the thought of money, as they develop an attitude toward their new job, may be partly at risk of increasingly attributing their own behavior to the money. These expatriates might even become metaphorically 'addicted' to the money rather than to more intrinsic features of the work itself (Solomon and Corbit, 1974). That may be partly why the more highly paid are often less satisfied than middle-income groups (Argyle, 1989).

In case this sounds a little far-fetched, let us remember that, although aid expatriates are often essentially meant to mentor themselves out of a job, the exact reverse sometimes happens. Of course, there are socio-economic reasons for this, such as being disadvantaged in terms of post-contract job security 'back home' (Harrod, 1974). However, in the South Pacific, for instance, Traynor and Watts (1992) describe how foreign expatriates jealously guard their positions against 'usurpment' by host counterparts, while in Malaŵi one expatriate had stayed for thirteen years beyond his original two-year contract period.

More generally, Littlewood speaks of expatriates' often 'marginal' status at home and the 'lure of the exotic' as the panacea for a range of woes and domestic disenchantments (1985, pp. 194–5). Dore has observed that expatriates, 'though often of mediocre attainments can live in a luxury unattainable at home [and] have a vested interest in inflating TC programs' (1994, p. 1432). Perhaps, then, they sometimes have an interest in *extending* these TC projects too?

Self-perception theory predicts that those who begin their contract with a monetary orientation may become increasingly reluctant to give up their post, even perhaps to the extent of pushing down on and discouraging their potential local replacements (see Chapter 9). Self-perception processes would thereby go part way toward explaining Kealey's (1989) finding that expatriates who went abroad largely for the money were also largely ineffective at technology transfer.

Some of Kealey's Canadians who said that they went for the money may have been *over*-relying on Theory X to account for their own actions. That is, one hopes that they would have had some more intrinsic motives as well! Just as low-paid 'volunteers' are often partly motivated by other concerns, like 'seeing the world' (O'Dwyer and Woodhouse, 1996), most aid expatriates could be expected to enjoy certain aspects of aid work in their own right. There is direct empirical evidence on the probable effects of paying people to perform an activity that they initially quite enjoy and would otherwise do for free. This kind of situation is also partly analogous to the comparatively high rates of pay enjoyed by expatriates working in developing economies. Hence, the research is doubly relevant to us.

The accumulated evidence from this research consistently indicates that over-payment is likely to undermine any initial work commitment. In the original study of this phenomenon, Deci (1975) paid student experimental participants to work on intellectual puzzles. These were normally so interesting that people would work on them for free. Following a paid session, however, Deci allowed participants the possibility of staying behind and interacting with the puzzle during their free time (which they would normally do unhesitatingly if they had not been paid). Length of time voluntarily spent staying behind during this free-choice period was Deci's operational definition of level of intrinsic work motivation. He found that the introduction of payment actually reduced the amount of time otherwise freely volunteered for work. In other words, (over?)payment had actually *reduced* work motivation.

This counterintuitive finding has since been replicated in a wide range of settings (Deci, 1987), including aid organizations and related work. In regard to the UN aid agencies, Harrod (1974) described a 'calculative staff involvement' (extracting maximal self-benefit from the organization, for minimal self-input) which was brought about by the constraint of 'remunerative compliance systems' (p. 198), or systems that relied on the 'carrot' of money to 'get things done'. Among Western health-care workers, the replacement of salaries by pay-for-performance actually reduced intrinsic motivation (Jordan, 1986). The reasons advanced for this, and for Deci's demotivation effect generally, include the idea that linking performance to payment begins to wrest the perceived control over one's actions away from 'internal' factors, toward 'external' factors. In the process of attempting to explain one's actions to oneself, overpayment insidiously undermines the feeling of being in control of those actions (Deci and Ryan, 1985).

In addition, delivering extrinsic rewards for performing an activity one already likes, by its essentially *impositional* character, may create a certain amount of ambivalence toward the work itself. Crano and Sivacek (1984) found that paying people to argue their own freely held position (regarding the decriminalization of marijuana) left them prone to switch allegiances when later exposed to any kind of counterpropaganda about the drug. As a result of over-reliance on financial inducement, people had become receptive to negative information, implying some kind of acquired ambivalence in their underlying attitudes.

Analogously, perhaps, expatriates working on international aid projects are bound to encounter negative events, such as organizational corruption or inefficiency, death, or disease. In addition, they will face the relative poverty of their much lower-paid host colleagues. In that (likely) event, the external weight of relative pay may encourage the development of ambivalent feelings toward the job, an 'acquired ambivalence' that seems to us worthy of future investigation.

Let us now consider those altruists who venture into aid work primarily to 'help others' (see Frisch and Gerrard, 1981; Snyder and Omoto, 1992). Here, and counterintuitively perhaps, the theoretical scope for demotivation by salary inequities broadens. The foundational model in this event is known as cognitive dissonance theory (Festinger, 1957). This is one of the most researched theories in social psychology (Reeve, 1992, p. 192; Wheeler *et al.*, 1978, p. 131).

Dissonance is synonymous with inconsistency, and Festinger's central idea is that people are often socialized, in Western cultures (see Chapter 10, this volume) to feel guilty about inconsistency among their 'cognitions', namely what they know they believe, feel, and do (see Chapter 2). For example, the cognitions (x) I am here to help restore social justice, and (y) I receive ten times more pay than my more needy local colleague doing the same work, would probably create dissonance (Adams and Rosenbaum, 1962; Austin and Walster, 1974). Formally, according to Festinger, cognitions x and y become dissonant whenever 'not x' follows from 'y'.

Festinger also argues that cognitive dissonance is emotionally painful, particularly if the issues involved (such as the issue of 'social equity', to an aid worker) are very important to the person concerned. Accordingly, the desire to reduce dissonance can be analogous, according at least to Festinger (1957), to the biological drive to reduce hunger. Littlewood has discussed (1985, pp. 195–6) the distressing nature of relatively sharp pay inequities between expatriate and host in Africa. One possible remedy for such distress is to change one's behavior to fit beliefs or feelings, such as working harder to justify or compensate for one's higher pay. That solution clearly becomes impractical, however, when one is earning ten or twenty times the local salary.

In such cases, according to Festinger, we may be obliged (1) to avoid the source of the discomfort physically; (2) to distort existing cognitions, so that they fit one another better; and/or (3) to invent or invoke new/other cognitions (Bersheid and Walster, 1967; Cohen, 1962). One form of cognitive distortion is disparagingly to Blame the Victim (Ryan, 1971), for example, for his or her lower pay (for example, 'They deserve their lower pay, they are lazy'). This would be another instance of the fundamental attribution error, described in Chapter 2. Invoking the cognition (z) 'People get what they deserve', effectively dilutes, and thereby reduces, the contradiction between x and y above (Festinger, 1957). The implicit Belief in a Just World (Lerner, 1970) reassures us that people deserve whatever they receive.

In international aid settings, Kealey (1989) has found that expatriates are often inclined not to mingle socially with their local counterparts, suggesting an effort simply to avoid the source of discomfort as much as possible. Blaming the victim is often increased whenever the observer feels that (s)he cannot help (Lerner, 1970), because it is precisely then that the victim provokes the greatest discomfort and dissonance (Lerner and Simmons, 1966). Unfortunately, to judge by our own experience, this feeling of being powerless to help is one of the very first feelings experienced when going to live and work in a developing country. As we also suggested in Chapter 2, the guilt-induced reaction of blaming the victim may be enhanced by the combination of expatriates' privilege and proximity in relation to their host counterparts. Perhaps these are partly why aid-funded consultants may end up derogating their trainees as 'unreliable, untrustworthy, incomprehensible, or unpredictable ... *feelings that are often mutual*' (Fox, 1994, p. 44, emphasis added).

Believing in a just world, and *social* justice in particular, is at the heart of one of the most influential and relevant theories of work motivation. In 'equity theory' (Adams, 1965), 'equity' means receiving outcomes (pay, benefits, recognition, etc.) in proportion to one's inputs (for example, effort, ability, qualifications). The theory postulates that people seek a balance between the ratio of their own outcomes to inputs and the ratio for other people. Thus, if a counterpart was earning half as much money as oneself and (s)he (or they) worked only half as hard, or brought half as much to the job, then a state of equity would exist between you and that counterpart. One would feel, according to Adams, a sense of social justice in the comparative salaries, and that would be comparatively satisfying rather than demotivating (Berkowitz *et al.*, 1987).

Like dissonance theory, equity theory proposes that imbalance between beliefs about input/output of yourself in relation to others is aversive, prompting an equity-restoration process. For example, you may be earning twice your colleague's wage despite being equally able, qualified, or hard-working. In this case as we have seen, equity can sometimes be achieved simply by changing one's behavior, say by working twice as hard as your underpaid colleague. In one workplace field experiment, Adams and Rosenbaum (1962) found that workers overpaid by the hour worked harder than their underpaid colleagues. The overpaid group was thereby able to act to restore equity with its lower-paid colleagues. Expatriates, as we have seen, are often perceived to be doing too much, and thereby neglecting their core mentoring function. As we have also seen, however, in the case of aid expatriates receiving ten to twenty times the local salary, such a maneuver would not only be impractical, but also probably impossible.

In the opening chapter, we saw how one expatriate took a radical step to resolve a similar problem, quitting work altogether. Along with the milder reaction of absenteeism, this solution has long been recognized as a possibility in the literature (for example, Makin et al., 1989; Patchen, 1959). Expatriate turnover of the kind illustrated in the opening Case Study (see Chapter 1) might therefore be partly mediated by equity-restoration processes.

Other expatriates may adopt the less radical solution of cognitive distortion (Lawler et al., 1968). This would probably take the form of psychologically inflating one's inputs to match one's outcome, and/or derogating the inputs of the lower paid to match their inferior outcomes (Adams, 1965). Greenberg and Ornstein (1983) found that workers initially reacted to over-entitlement by working harder, but this soon dropped below average, partly perhaps because they began to develop a sense of superiority (see Landy, 1989). A number of social commentators have also discussed the influence of the colonial scenario in distorting the psychology of the colonizer (for example, Fanon, 1985, 1991; Haley and X, 1987). These respected figures have each argued that colonial privileges create a sense of internalized superiority among the most well-meaning of altruists.

Even in relatively short-spell, experimental simulations, mock positions of privilege over prisoners (Zimbardo, 1982), students (Lerner and Simmons, 1966), and workers (Kipnis, 1972), have rapidly instigated victim blaming (Carr, 1995). This same tendency has been noted toward indigenous counterparts among Anglo-Australian and New Zealand communities (Lynskey et al., 1991; Marjoribanks and Jordan, 1986), as well as among expatriates working in Africa and the South Pacific (Carr, 1996b; Fox, 1994; MacLachlan, 1993b; Munro, 1996). Aboriginal Australians, for example, were seen as comparatively 'untrustworthy', 'money-grabbing', 'aggressive', 'dirty', 'unreliable', and 'unmotivated' (Marjoribanks and Jordan, 1986, p. 23).

In summary, we have seen that there is a good deal of culturally reliable and valid psychological evidence to indicate that people generally do not work for money alone. Those aid workers that do are liable to prove ineffective in the field.

The remaining majority, owing to the sheer magnitude of salary differentials, may be unable to restore a sense of self-respect simply by working harder or longer. For those who choose to stay on the job, both dissonance and equity theories predict that they will begin to feel a paradoxical combination of guilt coupled with superiority. Since the latter contradicts the former, expatriates may enter a form of self-deception, not necessarily being conscious of the ego-defensive maneuver that they have made (if they were, the defense mechanism simply would not function!). Any discovery of such paradoxical feelings among expatriates would, therefore, indicate TC expatriate demotivation as a result of comparative overpayment. To the extent that expatriates believe themselves 'superior', they are likely to wind down their effort on the job.

In a sample survey conducted recently at the University of Malaŵi, fifty-seven expatriate and host lecturers were asked to indicate how 'somebody like you' would reply to a number of statements about the terms and conditions of work in the university (Carr et al., 1998). Such indirect, scenario-type questions have been recommended for use with sensitive topics in developing countries, because they minimize any tendencies to give socially desirable rather than frank (but socially insensitive) answers (Sinha, 1989).

The results of this survey were as we have predicted. Expatriates were significantly more likely than Malaŵians to agree that expatriates felt guilty about their salaries, while simultaneously feeling that they were better employees. The respective items, along with item means, are presented in Table 6.1.

Although, on average, the internationally salaried expatriates tended to *disagree* that expatriates are better employees, the reader will note that the mean value is relatively close to the midpoint of the scale. This means that many would have agreed with the statement. Moreover, despite the use of scenario-scaling, there may still have been some 'faking good' and this would render the obtained mean an underestimate.

As yet, we have no direct evidence that such cognitions actually translate into demotivated rather than motivated behavior. Simultaneously, therefore, we have carried out a number of controlled experimental studies on the effects of pay differentials on *behavior*. These studies have been conducted in Australia, a 'Western' country, since expatriates do still largely originate from such settings.

Table 6.1 Items on which groups differed

Item	Expatriates	Malaŵians
Some expatriates on large salaries feel guilty because they earn much more than local workers	**3.4**	2.0
Expatriates are better employees than their local counterparts	**2.7**	1.6

Notes: Scale ranged from 1–5, with higher ratings indicating stronger agreement. Bold highlighted means are significantly higher than those that are not in bold.

Source: Carr et al. (1998).

In the first of these experimental studies, seventy volunteer Australian undergraduates experienced a variation of Deci's (1975) paradigm (Carr, McLoughlin, Hodgson, and MacLachlan, 1996). In addition to being paid varying amounts (AUS$1 vs. $2) for interacting with an otherwise intrinsically interesting puzzle (constructing three-dimensional geometric shapes from two-dimensional pictures), some participants were also informed how much money others were being paid for doing the same 'work'. In a baseline condition, some participants were paid nothing throughout the session, and this group tended to use the full free-choice period to interact freely with the puzzle – i.e., they were intrinsically motivated by the task.

By comparison, the introduction of payment alone (i.e., either $1 or $2, without knowledge of what pay others were receiving), partly analogous to overpayment perhaps, reduced the amount of free time volunteered on the puzzle. Moreover, compared with these paid subjects, those who also *knew* how much others were getting, whether it was less *or more* than their own pay, all quit working within a minute of completing the paid session, i.e., almost immediately. Several psychological processes have underlain this '*double* demotivation' process (see Chapters 9 and 12). While the underpaid may have felt anger, their overpaid counterparts may have felt *entitled* to refuse to work *unless* they were (over)paid, and/or guilty about being overpaid compared with others.

Although laboratory experiments like this are highly controlled, and thereby allow us to attribute causation fairly confidently to the payment variable manipulated, there is an ever-present risk that they do not represent the situation in the real world. This plainly crucial issue of external validity was addressed in a second, field experiment reported in the same paper. A total of 126 randomly sampled workers indicated their level of job satisfaction, as well as whether they would continue working even in the absence of financial necessity.

All of the respondents were carefully matched so that an equal proportion had reported being equitably paid and overpaid (as well as underpaid) compared with others doing the same sorts of work. Such pay discrepancies have become commonplace in Australia, as part of the advent of industrial reforms like enterprise bargaining and individual contracts. As elsewhere, they are often divisive and potentially undermining of social capital (Cox, 1995, pp. 46–7). In our study, those workers who reported being overpaid (as well as those who reported being underpaid) compared with their local counterparts were significantly less satisfied in their current jobs. Twice as many also (consistently) reported being ready to quit their current job if given the financial opportunity.

It is important to differentiate this study from other research, conducted in the United States and the United Kingdom, which has sometimes found a reaction of bravado (and possibly increased motivation) at being overpaid. Essentially, our study closely resembles Adams and Rosenbaum's (1962) experiment, in which participants were led to believe that a *social injustice* had occurred. There had ostensibly been a slip-up during placement, and the participants were told that they did not have 'nearly enough' (p. 162) qualifications and experience for the pay they were about to receive. In our study also, the 100 percent pay differential was deliberately introduced without 'reasonable' explanation, in an attempt to represent something of the social injustice of pay differentials occurring in developing countries (probably a conservative *under-*representation, because the order of magnitude in aid work, as we know, is frequently much higher).

In the UK, Adams (1965) found that workers reacted to a 10–15 percent overpayment with bravado, with no change in production. In the USA, Stepina and Perrewe (1987) explained apologetically to their laboratory participants that others had accidentally been receiving $3 rather than $2 (50 percent pay differential) due to some absent-mindedness about the proper university rate; again with no influence on production. Perry (1993) found that overpaid African Americans (earning $50 per week above the national median for occupational counterparts) were comparatively satisfied in their work, which Perry attributed to a sense of retribution for historical injustices against Black Americans.

Hence, in all those cases where bravado has been recorded, there were also very reasonable grounds, and therefore ready excuses, for it. In our opinion, without a sense of social injustice in the overpayment – without a sense of injustice being represented in the case scenarios – it is not surprising that there were no signs of demotivation. There was, after all, no impetus to begin to feel the guilt that equity and dissonance theories predict is necessary to precipitate cognitive distortion, or indeed demotivation of any kind.

Carr, McLoughlin *et al.* (1996) discuss a number of other methodological differences that possibly render the Malaŵian and Australian studies more representative of the issues arising in aid contexts. These include, for instance, the similar occupational status between our field study participants and expatriates working on international aid contracts. Of paramount importance, however, may be the degree of cultural (dis)similarity between expatriate and host. That is the crucial factor to which we now turn.

Potential applications

Remuneration

International salary allocations might be distributed among teams comprising both expatriates and host nationals. The general principle that economic resources can be shared among contributing employees has already been applied with some success to worker motivation in the West, where the technique is known as gainsharing (Aamodt, 1991; Florkowski and Schuster, 1992). In the cross-cultural setting of development work, such interlinked and interwoven salary structures might cultivate a much-needed sense of interconnectedness (Kealey, 1989), known as 'positive interdependence', among expatriates and hosts (Johnson and Johnson, 1990). This term denotes the sense that one person's achievement reflects on the others, a concept that has been suggested as a means of galvanizing workers into efficient and coordinated collectives (see Chapter 9). Outside of work, it could be useful to assess whether Kealey's (1989) 'voluntary segregation' applies equally to expatriates receiving local rather than international salaries. A reduction in segregation between hosts and *locally* salaried expatriates would indicate a link between group-based relative pay and sense of community.

A modest pay reduction with gainsharing would mean that locals would end up earning far more than their fellow countrymen not involved in aid work. Surely that would create major cultural problems? As a recent report concluded, distortions in local incentive structures undermine aid project successes (World Bank News, 1996). On the other hand, however, such incentives, and indeed aid projects themselves, already create severe inequities among local groups (for example, Kohnert, 1996; Monan and Porter, 1996).

In addition, some major aid organizations (for example, the United States Peace Corps, British Volunteer Service Overseas, and Australian Volunteers Abroad), as well as some Church Mission groups, and many universities in developing areas, already operate a policy of paying the equivalent of local salaries to expatriate workers. The very fact that they are still able to attract at least some good workers from overseas suggests the viability of reducing expatriate salaries. Such reductions might free up valuable resources that could be directed elsewhere. They would also go some way toward answering criticisms that there is a need for greater economy and restraint in aid organization salary budgets (for example, Dawkins, 1993; Dore, 1994; SBS, 1996). Finally, reducing salaries (already implemented before, see Harrod, 1974) might result in attracting fewer of those individuals who are initially motivated by financial rewards (MacLachlan and Carr, 1993a).

To sum up, our research, along with many other studies, elucidates how salary differentials may become demotivating. At this stage, we do not pretend to have answers to this complex problem; but do join others in calling for more discussion of it, as well as empirical research on it (Argyle, 1989; Jordan, 1986).

Selection

In a very broad sense, efforts to alter the financial incentives in the job are attempts to better fit the job to the person. An alternative pathway involves better fitting the person to the job – for example by improving basic selection procedures (Aamodt, 1991). The way that selection fundamentally works is by measuring people and identifying those traits, motives, etc., that actually discriminate between successful and unsuccessful employees. One then looks for those same characteristics among future recruits, using appropriate psychometric instruments to inform the selection decision-making process, in conjunction with interviews and biodata, etc. (for example, rural backgrounds for rural sojourns, see Denys, 1971; Parker and McEvoy, 1993). This 'psychometric' approach (see Chapter 7, this volume) is increasingly gaining ground in the expatriate literature (for example, Harrison *et al.*, 1996).

Conventional aid selection practice, however, relies on previous experience. This is despite arguments, and mounting evidence, that length of résumé and effectiveness in the field are not necessarily related, and may be *inversely* so (Gow, 1991; Kealey, 1989; Parker and McEvoy, 1993 [for example, 'Experience is what you get when you don't get what you want!']). Moreover, those attracted to service overseas are not always the most internationalist or sensitive in outlook (for example, Dore, 1994; Harrod, 1974; Jones, 1991). While there are some reports that personality factors do not predict socio-cultural *adjustment* particularly well (for example, Ward and Kennedy, 1992), Kealey found that successful technology *transfer*, from Canadian expatriates to local hosts in a variety of developing countries, was closely linked (negatively) with individual characteristics such as personal drive and ambition. Thus, personality factors may indeed predict field effectiveness among aid expatriates, and often in rather surprising or counterintuitive ways.

Is there any evidence linking overpayment and demotivation directly to personality? Research in the West has found a wide variety of reactions to the disclosure of pay differentials in a university setting (Manning and Avolio, 1985), indicating some kind of role for individual differences. Elsewhere, Lawler *et al.* (1968) found that altruists were more likely than non-altruists to react to overpayment by attempting to work harder.

In aid settings, where we have seen that differentials cannot simply be 'made up' by harder work, altruistically motivated individuals may thus actually be more inclined to distort their cognitions, because their guilt and discomfort is that much higher to begin with (Boucebci and Bensmail, 1982). There is, for instance, a correlation between religiosity and belief in a just world (Rubin and Peplau, 1973), which in turn has been linked to comparative derogation of people less fortunate than oneself (Rubin and Peplau, 1975).

Perhaps the most obvious personality factor suggested by our theoretical analysis is individual differences in *sensitivity to equity*. Indeed, in a laboratory setting, Huseman *et al.*, (1987), and King *et al.*, (1993) have found that greater 'equity sensitivity' predicted a reduced tolerance for both under- and over-reward. Such findings clearly suggest that equity sensitivity might exacerbate the demotivating influence of overpayment. Maybe, those who are initially most sensitive to social injustice may end up the most *handicapped by the international aid system*. As Culpin reportedly observed in 1953 (cited in Littlewood, 1985), the expatriates most negatively affected by their sojourn may be the idealists who desperately strive to fraternize with the local population/community. In the vast job-burnout literature, long pinpointed as being most 'at risk' is the highly committed human-services worker (Cherniss, 1980). In management literature Senge observes, 'Scratch the surface of most cynics and you find a frustrated idealist' (1992, p. 146).

In a replication of the experimental part of Carr *et al.*'s (1997) study, eighty-five participants completed a test of equity sensitivity before departing from the laboratory (McLoughlin and Carr, 1997). In this refinement of the original procedure, it was found that knowledge of others' pay and individual differences in equity sensitivity both contributed toward greater demotivation. To be precise, while 37 percent of the variation in motivation was accounted for by the pay-disclosure factor, a further 17 percent (which would be highly significant in terms of selection utility) was accounted for by individual differences in equity-sensitivity. When equity sensitive individuals were overpaid, they might become particularly sensitive to the money, refusing to work on a previously interesting job unless they continued to be (over)paid. More likely, however, within a relatively short experimental session, they felt the pinch of guilt more acutely than their equity insensitive counterparts.

The same experiment also measured people on the so-called 'Big 5' personality traits (Pryor, 1993). These are considered the traits possessed to a greater or lesser extent by all individuals. Their self-explanatory names are Extraversion, Conscientiousness, Openness (to new experiences), Agreeableness, and Neuroticism. In McLoughlin and Carr's study, scores on equity sensitivity were significantly predicted by a combination of Conscientiousness, Agreeableness, and Openness. These major traits may actually resemble Kealey's (1989) crucial traits of self-monitoring/adroitness, caring/other-centeredness, and low security orientation/active, which contributed positively toward TC expatriate field effectiveness.

The evidence, then, begins to indicate that the propensity to experience discomfort and demotivation at the experience of being overpaid is indeed measurable prior to being selected for aid work. Ongoing laboratory and field research is comparing levels of demotivation with scores on psychological tests such as belief in a just world and the OAI (Overseas Assignment Inventory) (Callahan, 1989). The latter in particular measures work motives, and has been designed and validated against the criterion of reducing unplanned expatriate turnover. In this task, use of the OAI has resulted in a reduction of up to 40 percent (Robbins et al., 1994, p. 724). Applied psychological research may therefore reinforce the idea that the effects of salary differentials can be ameliorated by selection, at least on the expatriate side (see Chapter 9 for a discussion of managing demotivation among host counterparts, and Chapter 12 for a systems perspective on managing the intergroup dynamics of salary differences).

Of course, it may be countered that the only way of recruiting people with the requisite skills is to offer no less in terms of pay and benefits than would otherwise be the case at home (Allen, 1987). This concurs with Cassen's (1994) advocation that aid expatriates should in future be recruited more economically from the developing world where the market price of a good professional is lower. In the meantime, however, and with regard to recruiting from the 'developed' countries, there would seem to be a need for recruitment research to be done, in order to determine whether higher pay and benefits actually do attract expatriates who are more effective in the field.

Socio-cultural and intergroup effects

The tendency to become demotivated by overpayment may also be amplified or dampened by group factors such as organizational and national culture. Australian culture traditionally stresses 'mateship' and 'egalitarianism' (Feather, 1994), emphases implying that the Australian workers (as in our studies) might be relatively prone to feel guilt rather than mere bravado at being overpaid. Similarly, however, Levine (1993) has found that Japanese employees felt more guilt than did employees in the United States, when being paid more than their fellow nationals doing the same job. Cultural effects therefore represent an important area for future applied research. Let us not make the same mistake that is sometimes made in the commercial sector, namely assuming that being a good technician at home automatically means being an effective manager abroad as if the cultural environment does not matter (Giacalone and Beard, 1994; Parker and McEvoy, 1993; Thomas, 1996; see also, Munandar, 1990).

With the increasing numbers of expatriates emanating from non-Western, for example East Asian nations, such cultural effects may become more germane to selection and placement decisions generally (Dore, 1994; Mamman, 1994, 1995). The tendency for researchers to overlook non-Western expatriates working in Third (and Fourth) World settings has recently been discussed by Carr *et al.* (1996). To return to our opening remarks about distinctiveness, among the most visible (and influential) characteristics of expatriates is their ethnicity and its degree of (dis)similarity to that of their hosts (for example, Herriot, 1991; Thomas and Nikora, 1996). Although ethnic differences constitute something of a 'hot potato' politically, Mamman (1995) has used this very fact to argue that they are likely to play a very significant part in determining intercultural effectiveness.

In that regard, Westerners' clear ethnic dissimilarity to some collectivistic aid hosts may be less of an issue than the degree of perceived similarity to the host of *non*-Western expatriates. Similarity is not always attractive (Fromkin, 1972; Gergen, 1994). There are a number of reasons for this partly counterintuitive prediction, which is graphically portrayed in Figure 6.1.

Social Identity Theory (see Chapter 9) predicts that groups are more likely to feel competitive toward comparable (i.e., similar) rather than dissimilar groups (Tajfel, 1978). Many 'traditional' societies that stress the collectivity before the individual (see Chapter 4, this volume) may also emphasize intergroup competitiveness, including sometimes rivalry and non-cooperation in the workplace (Munro, 1996). Earley (1993) found that largely individualistic US managers performed the same whether they were working alone or in any group context, whereas the more collectivistic Israeli and Chinese managers worked better when working with an ingroup (they became highly motivated) and *less* with an outgroup (they became uncooperative and demotivated). A corollary of inclusion is exclusion, or 'the dark side of community' (Cox, 1995, pp. 34–5) – a side of inter-community life that might be enhanced by shortages of natural resources,

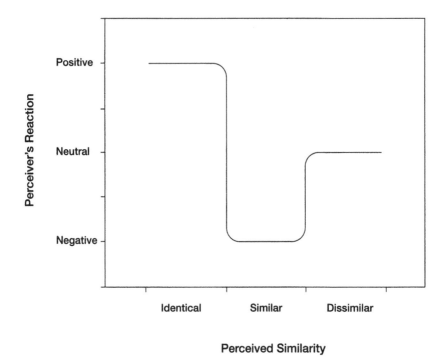

Figure 6.1 The 'inverse resonance' hypothesis
Source: adapted from Carr (1996b).

by propinquity, and/or by any 'outgroup homogeneity effect' (Augoustinos and Walker, 1995).

Intergroup competitiveness and non-cooperation, according to social identity theory, is motivated by group pride, and an extensive literature (for example, see Fisher *et al.*, 1981) indicates that receiving aid from a *similar* (rather than dissimilar) other can be symbolically threatening, and thereby perhaps more likely to become resented and/or criticized for its motives, or quality, or indeed sabotaged as a way of re-elevating one's esteem (Gergen and Gergen, 1971). Added to this, nationals of 'developing' countries may, as we have seen, internalize that 'developing' status, thereby perhaps making relatively negative attributions about the abilities of expatriates originating from elsewhere in that same 'developing' world (Eze, 1985; Mamman, 1995).

Anecdotally, we have described how African expatriate applicants to the University of Malaŵi were screened out, for whatever reason, despite their obvious suitability for the post, while Western expatriates, often with much weaker résumés, were not. In Papua New Guinea, non-Western expatriates had to work comparatively hard to earn acceptance and respect from students there. Carr, Ehiobuche *et al.* (1996) also discuss a number of experimental studies demonstrating that when people are made to feel more interdependent and collectivistic, their reaction to similarity becomes relatively negative.

One study involved forty-four host countries and over 2,000 US Peace Corps volunteers (Jones and Popper, 1972). It focused on what differentiated aid expatriates who completed assignments from those who did not, and concluded that 'cultural unusualness' (i.e., *dis*similarity) was a central factor. A similar finding, apparently involving both aid and commercial sectors, has been reported more recently (Parker and McEvoy, 1993). In each study, those expatriates who appeared more 'exotic' were also more likely to adjust generally and to complete their assignments. Recently, Carr, Rugimbana, and Walkom (1997) have found that Tanzanians tended to be more accepting toward expatriates originating from Western countries than toward expatriates from neighbouring countries.

From Figure 6.1, such findings may mean, *ceteris paribus*, that more similar ethnicity – depending on additional contextual and cultural factors – produces an 'inverse resonance' from collectivistic hosts. In applied mechanics, a small stimulus of similar frequency to a system will amplify vibration in and synergize that system. The small stimulus becomes a 'driving force' for much amplified vibration. What we are proposing is that in a *social* system, the exact opposite may occur. Perceived similarity and greater comparability may sometimes produce an amplified *negative* reaction. In that regard, we join Mamman (1994, 1995) in predicting that some visible ethnic features, as well as appropriate training in how to manage reactions to them, may partly influence who is likely to succeed versus fail in an expatriate assignment. Thus, a Cockney working in an inner-city community in Glasgow may in some contexts be less well received than a Fijian working in the same place.

There may also be links between demotivation by overpayment and the concept of inverse resonance. In Hong Kong, for instance, Japanese expatriates are paid more than their local counterparts, and they reportedly regard these local counterparts as second-class citizens, or *genchijin* (Wong, 1996, p. 98). In this particular cross-cultural context, expatriate managers may frequently Push Down (see Chapter 9) on their local colleagues, thereby disparaging and demotivating them. Yet, within wholly Japanese work settings, as we have seen, employees feel relatively guilty about receiving any more money than someone else doing the same job (Levine, 1993).

From Figure 6.1, one possible explanation for this slight discrepancy would be that Japanese expatriates resonate positively with other Japanese managers (ingroup) but resonate *inversely* with Hong Kong Chinese managers (outgroup). If so, receiving higher wages than one's local counterpart is perhaps less likely to prove guilt-provoking, and to that extent less likely to become demotivating. Depending on social context, therefore, double demotivation could be reduced, at least on the expatriate side. Of course, this might have a deleterious effect somewhere else in the system.

At the beginning of this chapter, we saw that increased recruitment of expatriates from developing countries has been recommended as the central plank for curbing escalating expatriate costs, a scheme that has already been partly implemented by UNDP (Cassen, 1994). Cautioning against this, however, Dore warns that, 'you may well have an Indian sent to Uganda and a Ugandan sent to India. Both live much better but are much less effective than they would be at home' (1994, p. 1432). Our concept of inverse resonance seems to support such a caution, and to that extent might eventually aid expatriate selection and placement. As others have aptly remarked, 'Seldom do we consider the possibility that a deficit in resources might result from some positive [communitarian] aspect of the culture' (Gergen and Gergen, 1971, p. 99).

Part 3

HOSTS ABROAD

7

WHO SHOULD ADAPT TO WHOM?

Like Zeffe in 'The case' (see Chapter 1), many managers and development workers from developing countries gain their professional qualifications overseas, often in donor countries as part of an aid project. From beginning to end, this can be a fairly stressful experience. Selection, for instance, can be a very competitive business, with a great deal of 'face' (see Chapter 4) at stake. This can be followed by the stresses of encountering a foreign country and its learning culture, often while separated from spouse and children. Finally, recipients are expected to return home and readapt to their old job, often as if they themselves had not changed in any significant ways. Thus, even after his return home, Zeffe's former sponsor(s) still expected him to be deferential toward an expatriate, even one who was actually his junior in many respects.

This issue of who should adapt to whom even arises at the very beginning of the process of sending hosts abroad. One approach, that which is usually adopted, involves attempting to select the candidate to match the project (for example, whose academic qualifications are most appropriate to a particular course). This approach, introduced in the previous chapter, is known as fitting the person to the job rather than vice versa, and has been explicitly advocated for students and visiting scholars from developing countries (Zheng and Berry, 1991). The alternative, much less frequently adopted, would involve somehow tailoring the project to fit available human resources. Much (but not all) of the present chapter expounds on this latter idea.

Over the years, organizational psychology has managed to strike a practical and reasonably effective balance between these two approaches, and in this chapter (as well as the next), we discuss the possibility of applying some of those techniques to aid hosts abroad. With the commercialization of higher education now in full swing, it is important that such techniques are considered (Howarth and Croudace, 1995). A recent review from the South Pacific found, for instance, that 38 percent of aid-funded students had not been completing their studies, particularly in those areas where human-resource shortages back home were most acute (Vallance, 1996).

Fitting the person to the project

Much of this research has focused on the ubiquitous employment selection interview, which remains the principal mechanism for making final decisions about who will get a particular job. Several very-readable reviews of the literature are available (for example, Anderson, 1992; Herriot, 1991), and, unless otherwise indicated, the findings summarized below derive from those sources. Typically, they reveal, selection interviews are 'unstructured', which means that they contain no standardized set of questions, or agreed upon good and bad answers (for example, Foster *et al.*, 1996). In our experience, aid-selection interviews are rarely different from this. Moghaddam and Taylor, for instance, report that selection procedures are sometimes erratic and ineffective (1987, p. 72). The central problem with such unstructured interviews is that they allow the intrusion of a whole range of selection biases.

Selection biases

In Chapter 2, we learned about the negativity bias, by which interviewers' impressions tend to be unduly influenced by minor slip-ups, that probably have little, if anything, to do with the candidate's ability to do the job in question. Some Westerners' tendency to make a 'fundamental attribution error' means that a range of negative dispositions are liable to be misattributed to any candidate who is unfortunate enough to be simply 'nervous' during the interview itself. As we have suggested, nervousness may be particularly likely in the case of selecting candidates for international aid scholarships and other, training-related, forms of aid funding.

A set of job-focused, benchmark answers to specific questions would probably reduce such biases, indicating that lack of structure may be the problem here. Indeed, more than 8 decades of research, and over 500 studies on the selection interview, have shown that *un*structured interviews generate unreliable assessments. Panelists' overall impressions of a candidate often correlate with each other as lowly as .30, and with future performance even less (logically, the predictive validity of any measure cannot exceed its reliability, because a measure can never correlate better with performance than with itself!). Evidently, then, a number of other biases may be entering into selection processes that remain unstructured.

The tendency to dismiss job candidates who make one slip-up during the interview implies that panellists are prone to make snap decisions. Indeed, supplying (versus denying) background information prior to an interview seems to have little or no impact on the final decision. It is as if interviewers largely ignore the candidate's history and record, much of which may be job relevant. This tendency to give undue weight to performance at interview, rather than, say, previous record, could assume a particular significance in aid-scholarship selections. These are often made by expatriate consultants, who have been especially flown in for the task. Added to this, candidates invariably make their own attempts to manage the impression they create. With all due respect to aid consultants' experience, therefore, they may still be relatively prone to some selection biases.

In Western organizations, snap decisions are also often made on the basis of essentially irrelevant verbal behavior. Research has found that any trace of a regional or foreign accent, or speech influency, or soft rather than assertive tone of voice, is liable to reduce the candidate's overall rating significantly. The same applies to less than 'assertive' non-verbal behavior (or body language), such as poor eye contact and smiling. Any and every one of these factors, we believe, is relatively likely to be encountered when Westerners are interviewing prospective students from developing countries (see Chapter 4, this volume). Humility, for example, is often the valued norm in these countries.

Keats (1993a) observes that overseas students who are in fact shy and reluctant to ask questions tend to be unfairly downgraded, while their counterparts, whose ability in that particular regard is 'better', tend to be over-rated. Through her own extensive research in an Asian–Australian context, this author has also found that English-language skills are often relatively uninfluential determinants of later success among aid-funded scholars (Keats, 1993b). Thus, Daroesman and Daroesman (1992) found that language difficulty 'did not correlate with success or failure rates', while the Australian government has advised that 'What is usually diagnosed in Australia ... as poor language skills is often [simply] inexperience in written expression' (DEET, 1991, p. 16).

Perceived similarity of others to oneself may have something to do with the halo effects (over-generalized good impressions) that apparently surround fluency in English, and this factor is widely recognized as a source of selection bias in Western organizations. In general, and in contrast to Figure 6.1 (see also Chapter 4), individualistic Westerners tend to prefer the company, and to like better, those whose beliefs and values are similar rather than dissimilar to their own (Byrne, 1971; Forgas, 1986). Such similarity reassures us by social means (see Chapter 3) that our view of the world is correct or valid (Festinger, 1950), and in an organizational context would, in the long run, act to promote (organizational) cultural uniformity rather than diversity.

Within interview settings, the tendency to prefer similarity over dissimilarity translates into a preference for candidates who are more similar to oneself on personality traits, attitudes, and perceived origins (for example, Gallois et al., 1992). In a team of expatriate fellowship consultants, and according to Festinger, it would be understandable if the similarity of the candidate became a comparatively salient feature of their decision-making processes. After all, finding someone with a similar outlook tends to be at its most reassuring (and alluring) when one is visiting 'exotic' places.

There are many more biases that we could mention, but is awareness training the simple solution? Research has shown that merely raising people's consciousness about such biases is not, in itself, sufficient to raise the quality of unstructured interviews. On the contrary, it may even create a potentially harmful *illusion* of genuine skill. This false confidence is interesting because it suggests that experienced interviewers, like newly 'experienced' trainees perhaps, may (sometimes) have an unjustifiable amount of confidence in their own abilities to make good decisions. In an aid context, Kealey (1989) has failed to find any link at all between expatriates' experience and their effectiveness in the field, while Gow (1991) has suggested that experience and actual performance are often related inversely. Thus, even trained and experienced expatriate selectors may, sometimes, be prone to bias.

Our brief sample of the potential sources of bias is by no means exhaustive, but fortunately these and other potentially distorting human factors can often be managed in a relatively simple and effective manner. The primary way of minimizing selection biases, in Western organizational settings, has been to provide selectors with some basic training in how to *structure* selection interviews. In the West, such structuring techniques have been found to increase inter-rater agreement from .30 to .80, with predictive power climbing from .14–.20 up to .62 (Anderson, 1992). If this could be shown to apply in educational selection, and with regard to aid scholars, then we would have the beginnings of a reasonable case for the application of these principles to aid fellowship selection.

With regard to the first possibility, we can examine selection of medical students. Here too there is evidence of selection biases of the kind described above, with ethnic and gender dissimilarity (for instance) often bringing an unfair disadvantage (Collier and Burke, 1986; McManus *et al.*, 1995). By contrast, structured ratings of prospective medical students have cohered together across panellists, as well as predicting who would fail and who would go on to achieve honors in their studies (Powis *et al.*, 1988). In psychology itself (Newstead, 1992), like other disciplines (Newstead, 1996), unstructured marking systems have sometimes proved unreliable. Structured selection tests, on the other hand, have helped to predict final degree results (Howarth and Croudace, 1995). Since selection biases may transfer from business to educational organizations, they might be *managed* in educational settings equally well.

Is there preliminary evidence that structured selection interviews can be appropriately applied to awarding aid scholarships and traineeships? In the past, taking Indonesia and Australia as an example, most decisions have been taken internally, namely within the host organization (Daroesman and Daroesman, 1992). More recently, however, the move toward greater aid partnership seems to have produced an increasing emphasis on *joint* selection, involving both hosts *and* donors. This particular aid partnership may, therefore, provide an indication of whether structured selection could work.

Structured interviews

There are several ways of structuring a selection interview, but possibly the most efficacious is the 'situational interview', so called because it involves asking the interviewee to describe what they would actually do in a series of known critical (to the job) situations (Robertson *et al.*, 1990). This type of interview is based on at least two principles. In consumer behavior, people predict their own behavior best in the here and now, i.e., if asked whether they would buy the product 'right now', rather than at some vague point 'in the future' (Sheatsley, 1983). In social psychology, the principle of correspondence or compatibility states that attitudes and behaviors must be equally specific (or general) to correlate with one another (Fishbein and Ajzen, 1975).

The procedure for designing a structured interview is relatively straight-forward and freely available (for example, Aamodt and Surrette, 1996). In one of the best and most common procedures, one begins by generating a short series of 'critical incidents'. These are just scenarios that would, typically, precipitate some form of crisis in the particular workplace (for example, a student becoming highly anxious about an assignment that is overdue). These situationally focused incidents can have a positive outcome, such as approaching the professor for help, or a negative outcome, such as literally failing by not seeking any form of counseling (Ballard, 1987).

Often, incidents like these will be garnered and screened by a panel of experienced employees and job experts, but in the case of educational aid to Indonesia there is already a research literature listing some critical themes. Studies of what discriminates successful from unsuccessful students in Australia have revealed, for instance, that readiness to cope with the deregulated and individualistic academic environment as well as being prepared to approach and question supervisors may both contribute toward final degree outcome (Ballard, 1987; Daroesman and Daroesman, 1992). This contrasts with Indonesians' experience in a Canadian context, where English-language proficiency may assume relatively more importance (DEET, 1991; Madill *et al.*, 1995). The point here is that each aid partnership will have its own 'situational interview'. The technique is context sensitive.

Having gathered together a set of ten or so positive and negative incidents, the panel will normally design the interview questions around these, later asking the interviewees what *they* would do in each critical situation. At the design stage, for each question, the panel decides what will count as 'poor', 'average', 'good', and 'excellent' answers. During the interview itself, each panelist will be given some such standard scale for rating each candidate on each (standard) question. If some factors are already known to be more predictive of success than others, then scores on the relevant item(s) may be weighted accordingly (Kinder, 1994). Whatever the case, and bearing in mind that an interview should never become inflexible (Herriot, 1991), the interview has now been structured.

A situational interview like this was recently designed, jointly by Indonesian and Australian consultants, to rank Indonesian academics for their current readiness to take up aid scholarships in Australia (Carr and Mansell, 1995). Although it remains to assess predictive validity against actual future performance, using the specially constructed Success Rate Scale (Daroesman and Daroesman, 1992), the preliminary signs were encouraging. For example, panelists' ratings correlated highly with one another, suggesting that the structured-interview scores were reliable. Also, the combined score for each question correlated highly with the grand total over all remaining items, indicating that each item was indeed tapping into a genuine overall 'readiness for overseas study', on which rankings had been based.

The overall cohesion among items was indeed high, with 'coefficient alpha' (the standard *index* of cohesion) being well above the minimum .80. Alpha is simply the average correlation across all possible randomly split halves of the test (i.e., all possible ways of dividing the test items into two sets). The logic here is that if all test items are indeed measuring the same general tendency, then whichever way the test is 'cut', there will be a correlation between halves. Performing this across all possible halves, and then averaging the correlation coefficients into a single alpha coefficient, simply renders the process less arbitrary and more thorough.

The high alpha in this partnership covered both psychosocial and critical intellectual skills (for example, analysis and synthesis skills, having a topic of study clearly in mind). Such bridging across social and intellectual domains lends considerable credence to the idea, suggested by the literature on this particular aid partnership and (some) others, that readiness for study can be theoretically constructed as a coherent factor, comprising both socio-emotional and cognitive elements.

In sum, methods are now available to tackle the complex and challenging task of managing aid scholarships. Being relatively easy to put into practice, such psychological techniques might, in the long run, provide very good 'value for money'.

Structured tests

'Psychometric tests', as they are known, fall into two basic categories, namely those measuring character traits and those measuring cognitive aptitudes. In cross-cultural contexts, and particularly those concerned with international aid, cognitive tests have so far proved to be fraught with potential biases and difficulties (see Chapter 1, this volume; Zindi, 1996).

This leaves us for the moment with tests measuring character traits. There, one possibility for further applied research would be to ascertain the predictive power of the so-called 'Big 5' personality dimensions (Costa and McCrae, 1988). As noted in Chapter 6, these consist of intro-/extroversion, emotional stability (neuroticism), agreeableness, conscientiousness, and openness (to new experience). Perhaps it is not unrealistic to expect that such traits, if verified locally and measured appropriately, might influence the outcome of an aid project involving study or training abroad.

In the interim, there is a growing literature on characteristic 'approaches to study', one that borders on, and has on occasion ventured into, 'developing' countries (for example, Richardson *et al.*, 1995). According to this Western-based literature, there are two fundamental study styles, namely 'surface' orientation, involving a rote approach to learning, and 'meaning' orientation, involving a pursuit of deeper understanding. The existence of a third factor, that of 'achievement', is somewhat more controversial (Richardson, 1994).

There also seems to be another questionable assumption within this literature, namely that a 'meaning' orientation, through its emphasis on 'critical' understanding, is the criterion toward which students should invariably adapt (for example, Baillie and Porter, 1996; Dore, 1994). As Dore points out, this view is not likely to be shared universally, for example among students with traditionally 'power distant' values (Chapter 14, this volume). Nonetheless, the approaches to study literature, and the system it represents, continue to imply that aid-fellowship students should be selected to have a more 'critical' attitude toward learning.

This attitude may indeed turn out to be somewhat predictive of academic performance among aid scholars studying at Western donor universities, but we shall question the criterion itself in our discussion of the return home (Chapter 8). In the meantime, however, there is a more basic psychometric issue, namely the extent to which instruments developed in the West are likely to be appropriate to take valid measures of study styles extant in more traditional societies. Are they capable of capturing indigenous approaches to learning, or might they be prone to create distortion and therefore prejudice?

Some preliminary warnings can be found in data recently gathered in Malawi (Carr, MacLachlan, Heathcote, and Heath, 1997), and Fiji (Richardson et al., 1995). In the Malawian study, we examined the correlations between items on a Western instrument, the Approaches to Study Inventory (ASI). We found both 'general' and 'specific' factors. Malawian students showed some evidence of a 'reproducing orientation' (general), although the students were generally motivated by non-material considerations (such as national development). There was also an 'organized study' factor, which has appeared in the West (general) but which, in the Malawian context of relatively regimented instruction (specific), may have reflected average ability and hence correlated negatively with grades.

The remaining factor, which we termed Achievement-in-Context, was predominantly motivational and specific. It reflected a strong desire to do well, thereby honoring one's elders (respect for authority) and family (respect for one's group). Consistent with the acute shortage of university places in the country, students tended to score at ceiling on this factor, a finding that we have reason to believe would generalize to other 'developing' countries. It has previously been found, for instance, in Kenya and Zimbabwe (Watkins et al., 1994), Indonesia (Emilia and Mulholland, 1991), Papua New Guinea (Wilson, 1987), and Fiji (Richardson et al., 1995).

The Fijian findings did not however prevent Richardson et al., from their Western perspective, concluding 'that the various forms of motivation are not guided by appropriate and effective [cognitive] approaches to learning' (1995, p. 426). Begging to differ, we have interpreted motivation to Achieve-in-

Context as a valuable autochthonous factor that should, in future, be measured more validly, through developing more locally relevant tests. Indigenous professionals have made the same plea in a study covering 44 countries and 740 tests (Oakland, 1995).

Our Malaŵian students also tended to prefer concrete and active modes of learning. In Western theory, that combination of preferences would represent the highest stage in a 'learning cycle' (Kolb, 1984). And in fact, those Malaŵian students who preferred concrete examples also tended to achieve a higher grade. Thus, ideas articulated overseas, as well as those developed at home, may *each* help to illuminate and optimize the study process.

Once such possibilities are taken more seriously, a psychological analysis may have much to offer in the full and proper assessment of characteristic approaches to study, as well as their role in aiding fellowships toward greater success.

Fitting the project to the person

The need for the *system* to adapt to *people* is widely recognized in organizational psychology. That is partly perhaps why contemporary textbooks on the subject tend to devote more chapters to 'staff development' than to personnel selection (for example, Aamodt, 1996). In the remainder of this chapter, therefore, we consider some of the principal ways in which organizations can adapt to people; in particular, those adaptations that may be apppropriate to aid scholars and trainees.

By now, there are numerous accounts and studies of the difficulties often faced by overseas students, many of them from 'developing' countries (for example, Barker *et al.*, 1991; Paige, 1990). In 1996, in the *Psychlit* database alone, there were more than 100 studies, many of them highlighting differences in cultural values (Hofstede, 1985), as discussed in Chapter 4. We cannot possibly review all of these studies here, which anyway would not do justice to the variety of contextual factors embedded within each and every particular aid partnership. Like the host universities themselves, however (Vittitow, 1983), these studies have tended to place the 'difficulties' in the individual rather than the system itself (Barker *et al.*, 1991; see also Guimond and Palmer, 1990). In that sense, past work has ironically suffered from a 'fundamental attribution error' (see Chapter 2), which is doubly ironic, given that 'it is comparatively rare for them [overseas students] to take these problems to a personal counselor' (Ballard, 1987, p. 110).

We now add our voices to the few who have argued for accommodating the *system* to the person (for example, Gopal, 1995; Moghaddam, 1993). But we shall do so differently, by presenting some of the techniques that have proved effective in managing people in organizations.

Community biases

Consider for a moment the profession of psychology. There seems no doubt that the curriculum in psychology, as presented to aid hosts studying abroad, is biased toward the West. From within the discipline itself there have been numerous critiques of the ethnocentric nature of much 'mainstream' psychology in general, and cross-cultural psychology in particular (see Chapter 1). Moghaddam (1996), for instance, describes how students originating from developing nations, many of them aid funded, continue to be served up forms of training that are heavily based on inappropriate assumptions and Western values (see also Rugimbana *et al.*, 1996).

One recent Australian study looked at perceptions and attitudes held by over-seas students from 'developed', as well as 'developing', East Asian and South Pacific countries (Carr, McKay, and Rugimbana, 1997). Preliminary analyses reveal that perceptions of prejudice among Australians at university generalized to the wider community, and for the purposes of this research these twin aspects were subsequently combined into a single index. Also rated were the quality of educational services and facilities, and whether students felt that they would recommend Australia to others on their return home. This issue has long been regarded as a crucial consideration with regard to future international relations and regional development (for example, Bochner and Wicks, 1972; Greenwood, 1974; Keats, 1969).

These visiting students tended toward equivocacy regarding both perceived prejudice and quality of service. This was particularly so among the full fee-paying students, indicating perhaps that aid-funded students felt less free to voice their concerns than their wealthier counterparts. Regardless of the source of funding, however, perceived prejudice and quality of service, unlike demographic factors, were closely linked to whether students would or would not recommend Australia to others on their return home. Feelings of discrimination and being undersold predominated in the students' open-ended comments to the (eco-nomic) survey as a whole. From Chapter 2, any negativity bias (Anderson, 1992), availability heuristic (Tversky and Kahneman, 1973), or negative contrast effects of being over-promised and under-delivered (Herriot, 1991) could leave these students feeling bitter. Full fee-paying students at least have the option of litigation or withdrawal of custom.

Some of these perceptions can probably be attributed to specific contextual factors. For example, in Australia, at the time, it was impossible for local students to 'buy' themselves a place in higher education. In a radically egalitarian and classless Australia, it may be considered somewhat 'un-Australian' to be able to do so. This could, understandably, trigger some resentment toward overseas stu-dents from their Australian counterparts, while the latter's individualism may also appear like prejudice to more socially oriented collectivists (see our systems analy-sis in Chapters 4 and 12). Interestingly, similar themes have been reported in relation to Australian Aboriginal students and the nursing profession (Goold, 1995; Tattam, 1996).

At a more practical level, we might draw an analogy with structured marking schemes, which we have seen often reduce bias when awarding grades (Newstead, 1992, 1996). Australian academics sometimes complain that 'foreign' students cannot write or indeed 'criticize' so readily as their Australian peers, and we have seen that such judgments may often be unfair. In that case, on the available evidence at least, structured marking schemes (like structured selection interviews) may reduce these biases. Certainly, they are standard procedure in organizational-performance appraisal (Aamodt, 1996). In the UK, too, most students in the higher education appear to favor them (Howarth and Croudace, 1995). Thus, we might consider the application of structuring techniques to the curriculum, in order to address any prejudice and disservice perceived to exist in that system (Newstead, 1996). We now review some techniques that could be so used.

Interview techniques

These tend to be semi-structured, and geared toward obtaining a very useful mixture of both quantitative and qualitative evidence. As such, they often go on to provide valuable design input to larger-scale, more tightly structured assessments and organizational interventions. In our own case at the University of Malaẁi, they proved to be very useful in realigning the curriculum toward both student and wider community needs.

The Critical Incidents technique was originally designed by Flanagan (1954). It involves asking respondents to recall critical positive and negative events from their own particular experiences, and as such is squarely client focused. In our case, in conjunction with the Save the Children Fund (SCF), we were aiming to train personnel to work in SCF's 'Children and War' project (MacLachlan and McAuliffe, 1993).

In Malaẁi at the time, there were many severely traumatized refugee children who had fled the civil war in neighboring Mozambique. What was needed was a course that would be relevant to aid-related work in Malaẁi's refugee camps. This was a setting with little history of professional counselling. In addition, there was no 'literature' to which such a course could be attached. Flanagan's technique was therefore applied in order to approach the issue of curriculum design from the point of view of students, some of whom had already had some experience of working in refugee camps.

These students were interviewed on site, in order to assess their own opinions about which incidents were actually 'critical'. Each student trainee was asked to relate two positive and two negative incidents, the details of which were briefly written down. In a standard manner, the students were then probed about what had led up to the incident, who was involved, and how it had been managed or mismanaged. A content-analysis procedure was then applied in order to identify and tally recurring, i.e., key, themes.

The technique did indeed produce a number of key requirements for working with traumatized children in a refugee camp. An example illustrates how the needs of future development workers can be radically different from, and passed over by, a Western curriculum. One of the striking themes to emerge from the critical incidents analysis was the need to help the children *forget* their traumatic experiences. Learning to live with trauma by distraction and dissociation may be a forerunner of the *tolerance* described in detail in Chapter 10. It certainly contrasts sharply with the Western emphasis on 'working through', and on 'dissonance reduction', described in Chapter 6, and it may be a highly adaptive strategy in the perpetually uncertain and insecure environments to be found in many refugee camps (Kanyangale and MacLachlan, 1995).

It would be easy to assume that these student counselors had taken a naïve approach to handling trauma, thereby failing to appreciate the importance of what Western therapists, theory, and dogma refer to as 'talking out' one's problems. But even if 'working through' and 'talking out' are better strategies in a Western setting, this might not be true in many non-Western contexts. It would also have been thoroughly at odds with the experiences of indigenous people themselves. Such issues must be thought through in the context of culturally appropriate curricula. Their merits and demerits could at least be considered by the potential counselors and practitioners themselves.

There is much evidence that critical incidents are also a practical option in Western countries. Aamodt and Whitcomb (1991), for instance, describe how the technique can be applied to conduct training needs analysis for university lecturers. In some of our own practical classes in Australia, we have found that (lack of) approachability is often a critical preoccupation of students there. If this is cause for concern among radically egalitarian and informal Australian students, then it may be more salient and intimidating still among those aid-funded students who traditionally value 'power distance' but also a nurturing relationship with elders (see Chapter 4). As we saw, it may be comparatively rare for an overseas student in difficulties to seek personal counsel from Western-oriented academics and services (Ballard, 1987).

The Nominal Group Technique (NGT) has been described by Sink (1983), and has been used in designing management training programs (Taffinder and Viedge, 1987). In Malaŵi, we applied it to the teaching of health promotion in areas such as HIV/AIDS prevention (MacLachlan, 1996c). In essence, the NGT is a participative vehicle that enables a small group to systematize any list of priorities with respect to some project, and it was felt that this might be particularly appropriate to development projects.

In our case, the 'project' in question consisted of isolating anti-health promotion ideas most likely to impede the communication of AIDS-prevention advice. An example of such disinformation would be a notion reportedly prevalent among students in the region at the time, namely that AIDS was an 'American Idea to Discourage Sex' (*Southern African Economist*, 1992). Accordingly, we investigated the applicability of the NGT to students and, moreover, their future courses on effective health promotion in 'developing' countries like Malaŵi.

Initially, the students were provided with the question, 'What ideas have you heard about HIV/AIDS which may encourage people to ignore health promotion efforts to prevent the spread of HIV/AIDS?' The first stage of the standardized NGT requires that participants write down as many ideas as possible in response to the initial probe. In a procedure designed to sidestep groupthink and other convergence processes (see Chapter 3), participants begin by individually noting down their own ideas.

Next, participants crafted and ordered their ideas into a table, selecting their own 'best' two and holding back two more in 'reserve'. Each member of the group in turn then publicly announced his or her two best options, which were displayed on a flipchart. If a 'best' idea is mentioned first by somebody else, a 'reserve' idea is simply substituted. Next, the facilitator (instructor) encourages the class to refine and streamline its ideas, for example, by collapsing several thoughts into one superordinate idea. By this stage, our group had generated fourteen different ideas, and these were then ranked for importance, in the standard way, by the individual participants themselves. Finally, the facilitator collected and compiled the rankings into a single, composite hierarchy. The entire procedure, undertaken with a group of final-year students, took just one and a half hours.

The belief that AIDS is just an American Idea to Discourage Sex was ranked overall fifth by these Malawian students. Other, potentially more damaging barriers identified through the NGT included perceived hopelessness and denial of the severity and finality of AIDS. Based on such considerations, we were able to teach several ways of attempting to combat counterpropaganda like the above, including an application of the NGT to solving as well as identifying the problem! From our perspective, however, the important point is that the technique proved useful in accommodating the curriculum *to local needs*. Recent evidence that the technique is valid comes from Venezuela, where the NGT has recently been employed to galvanize the community into self-help (Sánchez, 1996).

Overall, therefore, semi-structured social-assessment techniques show some promise with respect to educational development. Critical Incidents and Nominal Group techniques, just two of several possible examples, may be sufficiently flexible to bring the curriculum closer to the true requirements of students in (or from) developing countries. As we are about to see, such behavioral techniques may dovetail nicely with larger-scale, and even more structured, assessments.

Structured surveys

One of the most effective vehicles for accommodating an organization to an employee is also one of the most straightforward, namely surveying their opinions on how better to fit the workplace to staff-development requirements. As many social surveys do, our project to further develop the curriculum began with a comparatively small-scale, semi-structured analysis, using the SWOT analysis (MacLachlan and Carr, 1993b). This technique was first developed at the Harvard Business School (Thompson and Strickland, 1990), and is today commonly used in marketing management, which is appropriate to the marketing ethos that we have seen is now driving many Western universities. The acronym SWOT stands for Strengths, Weaknesses, Opportunities, and Threats. The SWOT *analysis* is intended to identify and articulate perceived strengths and weaknesses within the organization, as well as opportunities and threats from outside, in the market at large. In this case, we asked a small group of final-year students what *they* perceived to be the key features, in the Psychology Department and its curriculum, in relation to their future work and the community's requirements.

Although some aspects of the curriculum were perceived as Strong on relevance to developing Malaŵi, there was still too much Western psychology (Weakness), and an overall dependence on expatriate staff (Threat). The department was also seen to be Weak on selling the subject assertively to industry, despite the ongoing high demand (Opportunity) for managers trained in managing organizational behavior (Dubbey *et al.*, 1991). We therefore decided to try and reorientate our marketing strategy more toward locally relevant psychology, indigenous staff, and developing the curriculum in conjunction with industry and other prospective local employers.

These concerns provided us with a clear focus for the next stage of the project, namely a structured sample survey of both students and employers (Carr, 1994a). Following Thompson and Strickland (1990), we built on the reported strengths by narrowing curricular focus more on to local concerns, and then asked local employers for their opinions about its appropriateness to their own employment purposes. This meant that the curriculum was being taken *to* employers, i.e., being socially marketed, thereby addressing job market opportunities. In the event, feedback from employers was positive, meaning that it could subsequently be presented to pre- and reregistering students.

This information undoubtedly helped to increase student enrollments, which rose impressively over consecutive years. That increase in student numbers, and accompanying revenue to the department, enabled the hiring of several new Malaŵian members of staff, in the process reducing the threat of over-reliance on expatriates. Since these new Malaŵian lecturers were also recruited directly from local industry, that also enabled us to continue to develop a focus on local management issues.

Although some of the success in this project was undoubtedly due to the contextual factor of regional liberalization at the time (Bowa and MacLachlan, 1994), that same liberalization arguably belongs to the *Zeitgeist* of consumerism that has been passing through both educational aid and donor (as well as host) universities. Our experiences in Malaŵi may also, therefore, be relevant to developing better services for hosts abroad, studying at educational and training institutions in donor countries.

Concluding remarks

We do not wish to imply that students are invariably the best and only judge of what is best for themselves (see Abrami *et al.*, 1990). But currently, many aid scholars, and the projects that fund them, may be receiving a raw deal. First, increased consultation with the intended beneficiaries may help to improve the fit between curriculum and local needs. This would also provide a benchmark against which to measure the effectiveness of projects. That would enable selection techniques to be structured and honed toward selecting candidates with the maximum likelihood of success. Applying a few basic assessment procedures may therefore help to strike a more equitable balance between the needs of the system and those of the human beings within it.

8

TRANSACTIONAL POSITIONING

In 'The case' (see Chapter 1), we saw how Zeffe had returned to his home country on completion of his studies abroad. This is not always the outcome of educational aid and development projects, however, and over the years much has been written about the negative influences of so-called 'brain drain' (for example, Joyce and Hunt, 1982; Rao, 1979; Tan and Lipton, 1993; Vallance, 1996).

Perhaps this should not surprise us too much. In Malaŵi, for instance, newly qualified lecturers may be able to assist their families more by staying abroad (MacLachlan, 1996b). Instead of earning a salary of, say, £4,000 per year at home, by working in the United Kingdom they would net at least four times as much, enabling the same amount to be given to their extended family (who would expect to share in comparative fortune), with £12,000 left over for their immediate family. Thus, brain drain would benefit everyone within the newly qualified professional's own family, as well as constituting investment in the export of services and foreign-exchange income (Cassen, 1994).

In addition to these largely economic factors, there are also social influences that could be impacting significantly on brain drain. In developing countries, for instance, we have already seen how expatriate aid salaries are often the source of very harmful social injustices (see Chapters 1 and 6), while, on the other hand, working in exile may offer in some respects a more supportive social (and socio-political) environment (for example, Daroesman and Daroesman, 1992; Rao, 1979).

At the same time, however, 'the loss of trained staff is often a serious problem for institution-building projects', and may become more so with the call for more 'free-standing', market-driven fellowships (Cassen, 1994, p. 156). Clearly, it may often be desirable to manage brain drain more effectively than at present. In Fiji, for example, up to 40 percent of Australian aid-sponsored students have not returned home on completion of their studies (Vallance, 1996).

According to Rao's (1979) wide-ranging review, 'The reasons that prompt students to stay abroad . . . appear to be very similar in all developed countries [namely] if the influence of pull factors from abroad and push factors from home far exceeds the influence of push factors from abroad and pull factors from the home country' (p. 15). Rao's notion of a generic matrix of pushes and pulls is broadly reminiscent of the prisoner's dilemma game, which is widely referred to in development studies (Axelrod, 1984; Dore, 1994).

Internal exile?

As Dore (1994) has pointed out, however, in relation to prisoner's dilemma games in development studies, 'the calculus won't work' unless much of the human relations essence of development work is 'stripped out' from the model (p. 1428). Similarly, Carnevale (1995) has shown how non-Westerners often cooperate more readily than Westerners with fellow nationals when faced with such dilemmas. This may be partly why the clear majority of aid-funded students do indeed return (for example, Daroesman and Daroesman, 1992).

Such considerations suggest that we should 'accentuate the positive', i.e., the returnees, a little more, particularly since financial bonds and written pledges (as in Zeffe's case, see Chapter 1) are often effective ways of curtailing any exceptions to the rule (Cassen, 1994; Rao, 1979; but see Vallance, 1996). When we do, we begin to face up to a more substantial, *host-re-entry* issue, and one that seems to have been accorded far less attention than re-entry by *donor* expatriates (for example, Austin, 1988; Keats, 1969).

Such exile is a real and considerable prospect for any 'returnee.' In management and public admininstration, for instance, Western-style MBAs (Masters of Business Administration) and diplomas have been designed for another world (MacLachlan, 1993b, p. 156). In marketing, too, the curriculum is heavily Western, as marketing science has scarcely made any inroads into developing countries (Rugimbana *et al.*, 1996). As already noted, in psychology there has been a tendency for the curriculum to be driven exclusively by Western concerns (for example, Moghaddam, 1989, 1990, 1993; Moghaddam and Taylor, 1986, 1987; Nixon, 1994). As these many references imply, the situation in psychology may be magnified, and therefore 'high resolution', for our purposes of analysis.

The essence of these authors' arguments is that prospective psychologists funded to train in the industrialized nations are given a curriculum which is largely inappropriate, because it has essentially *evolved to meet the needs of industrially more developed countries.* Having invested themselves in such a curriculum, however, and as with the older generation passing on the old to the new, what hosts abroad develop is a competence, and vested interest, in professing that knowledge on their return home. In effect, their education prepares them for working only in the minority, 'modern' sector in their home country.

That constraint is in turn compounded by the individualistic and competitive ethos of Western higher education, which encourages rarified focus on a 'niche' rather than broad-ranging applied skills. The result, according to Moghaddam and Taylor, contributes toward a continuing 'dualism' between the poor majority and the elite minority. In effect, the aid process fails.

In Brazil, for instance, two major streams of interest to psychologists are, on the one hand, the exploration of highly individualized and internalized mental life through individual (often psychoanalytic) therapy, and, on the other, the empowerment of social groups through community participation. In some regards, these modes of therapeutic intervention are on an individual–community dimension, which parallels an 'economically elite – economically impoverished' dimension in Brazilian society. Different psychologies are thus being developed for different socio-economic groups within the same country, with psychology in some respects investing in, rather than helping to bridge, the chasm.

That conclusion may well generalize to other disciplines, as suggested by some of the comments made by returnees. They have long reported lack of recognition for specialized courses with a localized and applied (versus academic) focus; expressed a wish to retain links with the Western institution; and experienced difficulty in applying what they had learned (from Keats, 1969; to Daroesman and Daroesman, 1992). These problems may have something to do with the general flavor of Western curricula, which some would say claim to emphasize 'critical thinking' and theory rather than the equally valid 'learning by doing' (Jones and Barr, 1996; Kaul, 1989).

The AIDS Challenge game

The central conclusion of that discussion is that the educational product, to be effective, needs to *position* itself appropriately with respect to local rules of discourse. This rather abstract and disembodied concept is best brought to life by an example, which in this case is a game developed in collaboration with UNICEF to help Malawian schoolchildren learn about the dangers of Acquired Immune Deficiency Syndrome, or 'AIDS' (MacLachlan *et al.*, 1996).

The context for this study was the escalating risk to young people caused by seropositive 'sugar daddies' turning to younger women (to avoid higher-risk 'older' women), and in turn younger men turning to the more 'available' older women (Liomba, 1994). Prior to the project, and the reason for its introduction, 'didactic' health booklets had failed to improve AIDS knowledge to any significant degree (Nyirenda and Jere, 1991). In terms of game theory, which is a theory about the roles that people adopt and the way they communicate through these roles, the *wrong* game had previously been played.

Accurate knowledge of AIDS may be a necessary condition for behavior change, and it was in an attempt to raise knowledge levels to an acceptable minimum that we developed the AIDS Challenge game (Chimombo and MacLachlan, 1995). This is a modified version of 'Snakes and Ladders', in which secondary-school pupils move up the board by answering and explaining correctly various locally determined questions about HIV transmission and prevention. The game is designed to increase active involvement, which often enhances learning compared with passive receipt of information by reading (Lamm and Myers, 1978).

The results of this study were encouraging. After each of the four weekly sessions, students' knowledge scores improved incrementally and significantly, and there was an overall gain that was sustained over a two-month follow-up period. Moreover, compared with a control-group school, in which no pupils had been introduced to The AIDS Challenge game, those students who attended the test school but did not actually play the game *also* showed a slight but significant improvement.

We have discussed the possibility that these students had been the beneficiaries of a 'trickle-down' effect, suggesting that such games may have a wide range of application, for example, to communicating 'Facts for Life' (UNICEF *et al.*, 1993). In Malaŵi, however, the size of the effect was minimal and tentative. A more considerable trickle-down effect has since been reported from a South Pacific context (Gregory *et al.*, 1997). The possible systems dynamics of this process are represented in Figure 8.1, which is based on inverting the content of the archetypal 'Fix that Fails', to render an otherwise 'vicious' circle more 'virtuous'. This would be a fix that *works*.

Transactional positioning

Instead of simply interpreting the above findings as an argument in favor of games (and active fun) over more conventional methods of teaching, we prefer to see

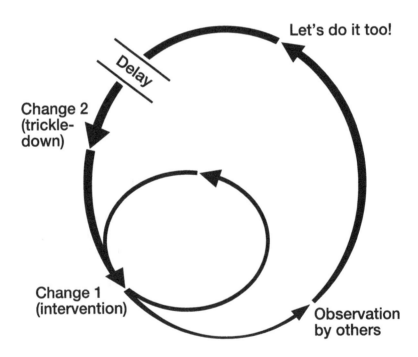

Figure 8.1 The possible dynamics of a trickle-down effect
Source: adapted from Senge (1992).

them as an example of playing the *right* game, for the particular educational project in mind. Sex is not normally the realm of teacher–pupil discourse. It is more easily the domain, perhaps, of peer relations, life, and fun. In a similar manner, perhaps, condoms might have been marketed more effectively in Malaŵi had they been packaged and promoted by association with fun rather than with clinical prevention of death and disease. An analogous proposal has recently been tested, and supported, in an Australian social marketing context (MacLachlan *et al.*, 1997a).

This study was based on a model of psychotherapy originally called 'transactional analysis', or TA (Berne, 1964), which has since gained credibility and popularity in management settings (Makin *et al.*, 1989), including training for aid extension workers (Bhattacharya, 1994; Hope and Timmel, 1995), and aid project managers themselves (Carr, 1996b). In at least one major company, for example, it has been evaluated as capable of improving intergroup communication (Nykodym *et al.*, 1991). It may therefore also be relevant to sustaining communication with, and among, aid organizations – especially with the growing tendency to view aid itself as a form of business transaction between donor and host (for example, Paul, 1996).

TA classifies relations into three broad categories or roles, namely parent, adult, and child. Briefly, Berne argued that effective communication requires both therapist (teacher) and client (student) to agree on their respective roles, whatever those might be. An adult-to-adult transaction is 'OK', provided each party wants that, just as a paternalistic, parent-to-child transaction is OK, provided each party wants that and can agree on who will be the 'child' (we all play the child from time to time). Such relationships, wherein each party is accepting of the role that the other is playing, Berne termed 'parallel', and they represent, according to him, the optimal social configuration for communication. By contrast, when an aspiring communicator attempts to play parent to a host who wishes to be treated as an adult, or alternatively an adult-to-adult posture is adopted when the pupil actively wants a nurturing parent, then the transaction becomes 'crossed', and communication will soon, according to Berne, begin to break down.

This prediction was in fact borne out in our study in an Australian university. The students rated two widely used AIDS-prevention advertisements, as well as the transaction being projected by the speaker in each, and their personally preferred transaction. As predicted by Berne's model, those students who perceived a crossed rather than parallel transaction also rated the advertisement as significantly less effective.

On the basis of this and other findings, we have suggested that basic transactional-type analysis of the relationship between aid partners may, sometimes, assist in the prevention of poor communication. In rural aid projects in Vietnam, for example, 'A major constraint is the paternalistic attitude of many extension workers' (Monan and Porter, 1996, p. 51); while in Bangladesh, the genuine (adult-to-adult?) involvement of local women has met with some success

(Meisner, 1996). By viewing aid therefore in terms of role relationships, a transactional perspective offers a conceptual framework for facilitating better communication between donors and hosts – i.e., one in which there is minimal role constraint.

Looking back on these studies, we see that Berne's model provides us with a tidy conceptual framework for interpreting the findings. In Malaŵi, MacLachlan *et al.*'s (1996) pupils may have preferred parent-to-child style transactions for conventional lessons, but child-to-child for talking about AIDS (remember that one is not literally a child, but simply relating to others in a child-like, perhaps fun-loving, way). Most of the older, Australian students, it turned out, wanted adult-to-adult transactions, whereas the commercials had projected a patronizing, essentially parent-to-child attitude. This attitude, it would seem from our research, was their downfall.

In itself then, a transactional analysis does not advocate any one role over any other. There is no insinuation that 'democratic', i.e., adult-to-adult, relationships are superior to say, parent-to-child. In those cultures that traditionally value paternalistic relationships between managers and workers (see Chapter 4), the important issue may be whether there is a parallel or crossed transaction, not the substance of the relationship itself (Salmen, 1995). In principle, at least, transactional-type analysis is relatively 'culture free'. Indeed, although relationships are always important to bargaining (Kniveton, 1989), they may be especially so for collectivistic groups (Leung and Wu, 1990).

In the hierarchical settings found in some developing countries, 'rationalistic' media campaigns, many of them aid sponsored, may be implicitly assuming that adult-to-adult transactions are appropriate. They might, however, be underestimating the potential of more paternalistic messages (provided they are delivered by community *in*siders, see Chapters 3 and 6). In Malaŵi, university students tended to perceive older family members (uncles and aunts, for instance) as much more credible sources of advice about HIV/AIDS than the national media (Carr, 1993), while nurses and doctors were consistently seen by undergraduates and school pupils as more credible than peers (McAuliffe, 1994). In egalitarian Australia, however, paternal advice may have reminded adolescents of the parents with whom they traditionally conflict, thereby generating a crossed transaction. Crossed transactions may thus consistently undermine the effectiveness of social-marketing projects, with what counts as an acceptable transaction being contextually determined. In cross-cultural settings, such as those provided by international aid, there is particular scope for different (and therefore crossed) role expectations, as we have argued in Chapter 4.

In one sense of course, this TA interpretation is almost too neat to be wholly true. In particular, TA is ethnocentric in restricting its transactions to roles based on the nuclear family, when the norm in host countries tends to be the extended family. The point here is that Berne's framework, at least as it stands, is too simplistic and rigid. Speaking more broadly, however, we may say that the concept of a transaction is consistent with that of a 'narrative', a construction that is more universally understood (Kashima, 1997), widely studied (for example, Bartlett, 1995; and our own 'scenario scales' for measuring double demotivation and motivational gravity, see Chapter 9), plus recognized as applicable to well-being (for example, Monk *et al.*, 1996; Murray, 1997).

One extremely flexible way of thinking about communication games has been suggested by Harré and Van Langenhove (1996). They have developed what is called a theory of 'positioning'. This involves the use of rhetorical (conversational discourse) devices, by which oneself and the speakers in a conversation position themselves in relation to each other. Positioning is about constructing implicit story-lines, or social scripts, from which the players play in concert (Goffman, 1959). This is known as 'first-order' positioning, and clearly resembles Berne's 'parallel' transacting. 'Second-order' positioning occurs whenever one of the players questions his or her part as cast by the other party, wishing it to be renegotiated. This is clearly analogous to Berne's 'crossed' transaction. In 'third-order' positioning, one steps back from the social-negotiation process, and comments on it, rather like this text at the moment.

As they stand at the moment, we find positioning theory more flexible than TA, but somewhat less accessible in its terminology. Perhaps, then, a balanced combination of the two terminologies could be struck, giving us a model that retained the flexibility of one with the practical utility of the other. This might also acknowledge its roots in TA-type thinking, whose key strength in the aid context, we believe, is its inherently social interactionist approach.

Our proposal is therefore for a 'transactional positioning' model, which would encompass (1) parallel positioning, wherein each party implicitly negotiates a *mutually acceptable* narrative for communicating with each other; (2) crossed positioning, where they cannot agree on mutually acceptable and interlocking roles; and (3) meta-positioning, when (1) and (2) are discussed at one step removed. We also envisage the possibility (see Part 4) of plural positioning, whereby the same players may alternate between (1) and (2) to suit situational contingencies, for example, at work versus socially.

How can this help us better to understand the situation of the internal exile? We can begin to answer this question by briefly reviewing a number of (we believe) fairly common scenarios, or 'narratives', in developing countries. These include expatriates becoming frustrated by what they see as broken promises, poor timekeeping, and passive obstruction on the part of host colleagues. Adopting a meta-position, however, these could often be cases of cross-positioning, whereby hosts are essentially resisting, implicitly perhaps, the expatriate's implicit attempt to play out an unwanted, inherently patronizing narrative (Fox, 1994).

In a broader sense, maybe we can think of set curricula as also being a variety of attempted positioning. When Dore (1994) describes the hostility generated by his suggestion that a Western-style of education (see Chapter 7) is best, that undertow could be attributed to cross-positioning and sociocultural resistance. When Moghaddam and Taylor (1986, 1987) describe 'dual growth' (or social division) within developing countries, perhaps returnees have adopted a parallel position with respect to the Western institution abroad and the modernized sector at home but a crossed position with respect to the majority traditional sector, where aid is often needed most?

Thinking about such cultural and intercultural relations in terms of transactional positioning may eventually help to create a better 'fit' between educational product and its consumer from a developing country. The first step in any such fitting process would be to attempt to measure some of the social constructions that would be acceptable to ('parallel' for) each party in the aid partnership. For example, we have discussed how workplace relations may, depending on context, be guided by various cultural narratives, such as neo-Confucianism (Carr and MacLachlan, 1997). In clinical practice too (where Berne's ideas originated), 'narrative therapy' seems to be gaining ground (Monk et al., 1996). A next step might be to test whether aid projects that position themselves parallel to their hosts are more effective than those that become crossed. Elsewhere, for instance, we have discussed how different academic subjects may resonate differently with different audiences from developed versus developing countries (Carr and MacLachlan, in press).

Perhaps the most obvious way to increase the likelihood of parallel positioning, in educational development projects, would be to conduct all training in-house, and at home (Moghaddam and Taylor, 1987). In reality, however, this may be a long time coming (Cassen, 1994). In Iran, for instance, more than 90 percent of completing postgraduates are reported to have been trained in the West (Moghaddam, 1993). Indeed, in many developing countries, the norm is still to send postgraduates overseas, either to 'Second World' countries such as Australia and Canada, or to the former colonial powers, such as Britain. Evidently, then, the possibility of cross-positioning remains, often reinforced by the history of 'communication' (colonialism) between donor and host.

As well as the approach described in Chapter 7, one interim tactic might be to supplement training programs with an extended debriefing course (MacLachlan, 1992), just as many existing programs already have an induction course. Debriefing is recognized to be an important component of industrial training programs, and yet African students in Britain (for instance) are not taught how to translate the principles they have learned into practical action at home. We have seen, too, how students returning to Asia often felt like they had been thrown (back) in at the proverbial deep end.

The temptation here, for the host returning from study abroad, may be simply to teach what they themselves have been taught. This, of course, is perfectly reasonable; otherwise, what was the point of getting the qualification in the first place? Educational and training institutions may thereby become infused with a 'replication mentality'. In debriefing, however, the about-to-graduate trainee is guided through the ways in which what has been learned may actually be applied. Thus, the debrief is the point to which all prior activity has led. It is where competent trainers could start (some of) their real work.

Pathways to a psychology for development

How might this work be structured? In our own attempts to contribute to health and welfare and organizational issues, through the development of psychology that is relevant to development issues, we have extrapolated and described some basic principles. They are presented now in the belief that they may have some generic value, to disciplines and projects other than our own (for more detail, see Carr, MacLachlan, and Schultz, 1995; MacLachlan and Carr, 1994).

Realization

This is the aid equivalent of good old-fashioned powers of observation, though not as easy to do from within a Westernized perspective as it may seem. In this case, debriefees might be taught to meta-position themselves with respect to the business of aid, effectively asking: 'Could this development project or practice be ill conceived from a psychological (or sociological, or economical, or whatever) point of view?' In our case, for example, we have described in Chapter 2 how donation appeals may be ill considered from a social-psychological point of view. In the current chapter, too, we have seen how academic cultures may be cross-positioned with respect to their own students.

Rejuvenation

Research and theory in social science are often temporarily discarded, as they fall out of vogue. Until they come back into vogue, however, they may be useful in other cultures and contexts that are currently facing a different set of development problems. We saw, for example, in Chapter 3 how some of the classic research on communication networks could conceivably be rejuvenated by adapting it for applications to the social dynamics of aid decision-making by donors. We have also learned, in this chapter, how AIDS knowledge may be a comparatively outmoded research topic in some Western settings such as Australia (where technology has recently helped disseminate information to near ceiling levels), but may still be a necessary condition for Malawian schoolchildren to make HIV prevention possible.

Refutation

This pathway entails considering the possibility that an 'established' principle might not actually generalize to a particular development setting. In Chapter 4 we learned that value patterns may differ greatly across cultures. A Western 'participative' technique, such as Management by Objectives (MbO or 360-degree feedback), may work quite well in some circumstances, but may still be wholly inappropriate in a societal or organizational setting in which hierarchy and certainty are valued. The present chapter, too, mounts a challenge to the position that Dore had once simply assumed, namely that one type of schooling is always the most efficient path to economic growth (1994, p. 1431).

Reconstitution

This is analogous to the process of alchemy, or combining a collection of otherwise 'base' elements into a valuable compound. In Chapter 5, for example, we learned how various elements from social psychology and psychophysics can be combined into a formula for more effective development projects. In the current chapter we have tried to combine elements from various theories, to suggest that to approach development from a more dynamic and interactive 'social games perspective' may be valuable too.

Restatement

This pathway involves developing *qualifiers* for existing principles, clauses that acknowledge that behavior may in fact be motivated by a plurality of intercultural factors. In Chapter 6, for instance, we saw how the influence of economic rationalism may be tempered or even compromised, psychologically speaking, by the impact of expatriate–host salary gaps. In the present chapter's terms, expatriates and their hosts may become implicitly cross-positioned.

Reflection

This is the reflective and evaluative process characteristic of debriefing, by which many observations are eventually seen to be connected by a unifying thread or principle. Perhaps the unifying theme in Chapter 7, and one that runs throughout most social and organizational psychology, is that genuine participation (of some kind) is preferable to, and more sustainable than, didacticism and constraint (as defined by the 'recipient'). That theme, of course, is implicit in the concept of transactional positioning, and is discussed as one of the linchpins for this book as a whole in Chapter 12. Thus, it may be that the more one attempts to cast an aid recipient into an unwanted or irrelevant role, the less one can effect the project.

Concluding remark

As we proceed with the book, readers may begin to see these pathways unfurling in other studies and ideas. You may also begin to see other pathways, or *combinations* of them, that may be applied to the same project or problem. In the next chapter, for example, we describe how McClelland's N. Ach (see Chapter 1), can be subjected to our 'pathways analysis'. Glancing back at Chapter 6, we examined intergroup relations between non-Western expatriates and their non-Western hosts. In the original derivation of that particular example, all six pathways were applied to the same (inverse resonance) prediction simultaneously (details are contained in Carr, Ehiobuche *et al.*, 1996)! Above all, then, pathways analysis has the potential to be both process oriented and pluralistic.

Part 4

HOSTS AT HOME

9

INTERCULTURAL WORK
DYNAMICS

In Chapters 1 and 6, we saw that technical cooperation projects, or TC, sometimes serve to highlight rather than decrease the gap between rich and poor, not only between (1) donor and host, but also (2) between host and host. The corollary of privilege–proximity, of course, is that someone else is relatively deprived. In his sociological treatise on envy, Schoeck (1969) describes both righteous indignation at the clear good fortune of others (which may be especially noticeable to those with little by way of material wealth, see Chapter 5, this volume), as well as envy that may be directed at those who are more socially comparable (Festinger, 1954). Envy today is less of a repressed topic in management generally (Bedeian, 1995), and the recent development-studies literature warns that expatriate salaries often create indignation among hosts (Dore, 1994), while aid benefits may precipitate witchcraft between them (Kohnert, 1996).

Perhaps indignation and envy, then, are partly behind complaints that host counterparts in TC often become too busy with their own job to spend time on the aid project, or even leave for another job; and why up to one-third of agricultural-research projects have been found faulted by inadequate local salaries, or even shortages of local personnel (Cassen, 1994). Such complaints, as Cassen observes, have not been investigated systematically, and in this chapter we investigate the issue with an 'intercultural' approach (Krewer and Jahoda, 1995).

That should not be confused with a *cross*-cultural analysis, which makes comparisons across cultures, or with a *cultural* analysis, which examines cultural behavior within its own unique terms. An intercultural perspective focuses on the various cultural influences that may be operating *within the same psyche at any one time* (technically, this could also be referred to as an 'intracultural' perspective, but to avoid confusion we will use the same terminology as others have done prior to us). In a South Pacific context, bilingual people who read and answered questions on a moral dilemma in their traditional language offered more traditional, collectivistic solutions, whereas reading and responding in English resulted in more 'modern', individualistic suggestions (Taylor and Yavalana-vanua, 1997). Other examples would include the behavior of second-generation migrants, who experience a mix of traditional and modern values, or a Western- and perhaps aid-trained manager working in an African organization. Both groups are liable to experience various stresses and strains, as they struggle to cope with often conflicting values and motives. It is easy to see that the intercultural approach is inherently 'dynamic', as well as inherently applicable to *work-*places inside developing countries.

There are at least two basic ways for these dynamics to unfold, namely in concert or in competition. In one scenario, the various cultural values may each motivate the person in the same direction. In 'The case' (see Chapter 1), some of the resignations may have been motivated both by a Western-style training in workplace equity, and traditional values concerning face, a possibility that we explore both theoretically and practically. In a second scenario, various cultural values might push and pull in completely opposite directions. Traditional collectivist values might clash with those in aid work via managerial training that stresses individual achievement, or the competitive ethos of aid organizations locally, or through the dilemma of allocating resources to an outgroup at the expense of an equally needy ingroup.

Such dilemmas have been succinctly described by a former Dean of Makerere University as being 'caught up between two moral worlds' (Mazrui, 1980, p. 119). Schoeck went further, describing developing countries as suffering from a 'negative selection' (1969, p. 345). A possible example of the latter (negative placement) comes from Papua New Guinea, where Bau and Dyck (1992) found that subordinates who scored highest on entry tests of ability were later ranked lowest by superiors for officer selection, as if the superiors were acting jealously to protect their own positions from potential threats.

Schoeck views envy as a primary and inevitable barrier to development, but our review of recent theory and evidence will suggest that it is often secondary to higher, socio-cultural values – in short, that it reflects not a weakness but a potential strength. We explore the development of culturally appropriate group techniques, as well as traditional metaphors that might be factored into aid training projects, adopted by aid organizations in their overseas offices, and utilized when implementing aid projects at the local level.

Indignation (at expatriate salaries)

Reducing input

As we learned in Chapter 6, pay is less often the cause of motivation than demotivation (Herzberg, 1966). It is also recognized in the sociological literature (Schoeck, 1969), as well as in social psychology (Argyle, 1989), that dissatisfaction with wages turns less on pay itself than on *relative* pay. Adams's (1965) equity theory predicts that workers will often react to perceived comparative underpayment by reducing their inputs to match their comparatively low outcomes.

Laboratory and field research has since confirmed that prediction. In the original study, workers who were underpaid in terms of hourly rates reacted by reducing the quantity of work completed, while pieceworkers who were underpaid increased the number of units they completed (simply to raise their pay) but simultaneously reduced quality, as if to secure a form of revenge on their employers (Adams and Rosenbaum, 1962). Both groups therefore reacted within their means to lower their inputs, with the intention of restoring social justice to their pay situation.

In addition to replicating these findings, more recent studies have found that input can also be reduced by forms of industrial sabotage, such as theft (Greenberg, 1990). Other, more commonly used tactics include absenteeism (Schwarzwald *et al.*, 1992), and even turnover (Summers and Hendrix, 1991). These in their turn have each been interpreted as forms of symbolic withdrawal (Schmitt and Merwell, 1972; Valenzi and Andrews, 1971).

In the context of aid work we have already seen, in Chapter 6, that locals and expatriates may tend not to mingle socially (Kealey, 1989). In addition, MacLachlan and Carr (1997) have suggested that pay-related conflicts may more easily be played out at work itself, and in a form far more subtle than the outright resignations in 'The case' (see Chapter 1, this volume). At work, social resistance may be manifested by a plethora of inefficiencies, demotivations, and distractions, many of which need not be consciously undertaken. Similarly, perhaps, Trompenaars has described how hosts may develop 'their own local standards which become the basis of their solidarity and resistance to centralised edicts . . . [As] soon as the attention of head office is diverted to other matters, normal life proceeds' (Trompenaars, 1993, pp. 41–2).

To outsiders such as aid expatriates, these actions may give the appearance of dispositional problems, reflecting simply a *lack* of motivation. In reality, however, we believe that the very opposite is often occurring. An apparent lethargy could well be masking an active and dynamic, intercultural resistance, of the kind perhaps that has recently been reported regarding the rapidly developing Indonesian automotive industry (Harriss, 1995).

Feelings of inequity may even impact on hosts' decisions about whether to return or to stay in external or internal exile, after earning their professional qualifications in the donor country (see Chapters 7 and 8). In more general terms, organizational commitment, and motivation levels generally, are often reported to be relatively low in developing countries. In Nigeria, for instance, the term 'Not-on-Seat' refers to workers who have clocked in but are actually absent (Munene, 1995). In Zimbabwe, the equivalent term is 'Jacket People', because their jacket is on its chair even though its owner is not (Munro, 1996). In Malawi, an equivalent term exists, namely 'Temporary Out'.

More direct evidence of demotivation is supplied in Carr, Chipande, and MacLachlan, one of our previously reported studies (1998). We surveyed host professional workers at the University of Malawi, concerning their feelings about salary differentials at their workplace. These hosts at home were being paid at the time only 10–20 percent of the wages received by their visiting expatriate counterparts, on aid contracts. Both groups were asked to estimate the level of fairness in, and demotivation caused by, the expatriate–host salary differences.

Malaŵians tended to report a greater sense of unfairness, as well as making significantly higher estimates of the demotivation being caused among host employees generally (in Fechner's terms, see Chapter 5, this volume, someone who has relatively little may be more likely to notice, and to feel, even the slighest differential in pay). The precise items, and their means, are presented in Table 9.1.

From Table 9.1, the Malaŵian lecturers were also statistically more likely to report being demotivated than their aid-salaried colleagues. The latter, however, were less likely than both their fellow (predominantly Western) expatriates on local salaries *and* the Malaŵians to perceive that there was indeed a problem with host demotivation. The Malaŵian lecturers also differed from the expatriates in advocating that pay and conditions should be, wherever possible, identical and fair. The expatriates on local salary tended to 'side' on these two items with their fellow expatriates, suggesting perhaps that they did not see this as a practical option (see Chapter 6 for more discussion).

However supportive of the notion of social justice these findings may be, they do not tell us directly how those feelings translated into actual workplace behavior. Also, we cannot tease out the potentially confounding effects of pay comparisons made by expatriates *among themselves*. Some of these were earning twice the salary of others representing different aid organizations, a situation that seems to be widespread (for example, World Bank News, 1996). In addition, the various recruitment strategies of aid agencies may have attracted expatriates of varying calibers and motives.

As described in Chapter 6, the issue of perceived unreasonable underpayment was also investigated under experimental conditions. Underpaying students for working on an otherwise intrinsically motivating task later resulted in a tendency

Table 9.1 Items on which groups differed

Item	Expat (*aid salaries*)	Expat (*local salaries*)	Malaŵian
How motivated are you to achieve the objectives of your job?	**4.6**	4.5	**3.9**
Local people are demotivated by the large salaries that some expatriates earn	**2.9**	**4.4**	**4.2**
Expatriates who work abroad should work under the same terms and conditions as local people	2.2	2.7	**4.1**
Most companies are unfair to their local employees	3.3	3.3	**4.6**

Notes:
Scale ranged from 1–5, with higher ratings indicating stronger agreement. Bold versus unbold means differ significantly from each other.

Source: Carr *et al.* (1998).

to quit the work within seconds of a free-choice period commencing. Individuals who did exactly the same work for exactly the same money, but without the knowledge that they were being underpaid, did not become so demotivated. In a replicating study of actual employees, underpaid workers were evidently angered by the social injustice of their lower pay, and reacted by drastically curtailing their previous goodwill and commitment.

While the Western (Australian) location for this research obviously makes us wary of generalizing too readily, there clearly are *potential* parallels with hosts at home. From the University of Papua New Guinea, for instance, Marai (1997) describes a physical fight between host and expatriate lecturers, one whose surface cause was demarcation about who would teach which specialist applied courses, but whose flames may have been fanned by salary gaps. This confrontation was very consequential. As a result of it, the expatriate lecturers, rather like the Tanzanian managers in Chapter 1, left the organization completely.

Need sensitivity

The issue of generalizability across cultures clearly remains crucial. Opinions concerning the cross-cultural applicability of equity theory varies widely, however. At one extreme, it has been suggested that 'in all sorts of societies . . . interactions have to be equivalent for both parties' (Makin *et al.*, 1989, p. 46). This view links the notion of equity to a widespread 'reciprocity norm' (for example, Deaux and Wrightsman, 1984, p. 228). At the other extreme, some have implied that the calculative aspect that underlies equity theory may be largely absent in many non-Western cultures (Owusu-Bempah and Howitt, 1995). The motive to eliminate cognitive inconsistency (on which equity theory is partly based) might not apply in many non-Western settings, although we recognize that the process of equity restoration may be driven more by the desire for social justice than for personal consistency.

Fortunately, the empirical evidence itself is a lot clearer. As well as hypothetical scenarios in which monetary resources must be distributed among characters who are variously industrious and needy, the research has also tended to involve university students, from both Western and non-Western countries. University students from developing countries are similarly likely to be reasonably comparable (Chapter 7) with university-educated 'professionals' working as hosts at home, on international aid projects. Furthermore, the major alternative *principles* envisaged for the operation of 'distributive' social justice include not only equity and need, but also equality, i.e., equal distribution of rewards, regardless of individual input. All things considered, therefore, the experimental evidence is reasonably relevant to the kind of situation in which we are interested.

The studies in question use a common methodology. The participants are given a number of scenarios or vignettes, each describing some available resources, usually money, and some potential recipients, often other workers of some kind. The participants' task is then to decide how much of the money or other resources should be apportioned to various kinds of worker. These tend to be of three basic kinds: the harder-working employee (equity); the more needy employee; or, neither of these two, with resources being allocated equally to both regardless of particular input (egality).

In one study (Berman *et al.*, 1985), the most popular option among (49 percent of) Americans was equity, while most Indians (52 percent) chose need. Nonetheless, 16 percent of Indians still chose equity, 16 percent of Americans still chose need, and equality was ranked intermediate in both groups. Moreover, in deciding how to *cut* resources, *both* cultural groups most commonly opted for need as the prime consideration. More evidence of overlap across cultures has been reported by Marin (1985). Although Indonesians considered need more than Americans did, there were no cross-cultural differences concerning preference for equity over equality. Radley and Kennedy (1992) remind us that the principle of 'equity' is not the sole property of the more educationally advantaged in the community, while Fijneman *et al.* (1996) have found that people in countries as far apart, both geographically and culturally, as the United States and Turkey often expect to receive as much out of a relationship as they put into it. In short, there appears to be considerable overlap between cultures regarding the rules of distributive social justice.

One particular concern may be highly relevant to (aid) workers from collectivist backgrounds, namely the social categorization in- or outgroup. When dealing with an outgroup, and consistent perhaps with the concept of inverse resonance (see Chapter 6), Leung and Bond (1984) found that the collectivistic Chinese adhered *more closely* to the principle of equity than did the more individualistic Americans. Similarly, in a resource-constrained scenario, Hui *et al.* (1991) found that the underpaid Chinese believed more than the Americans in rewards being distributed in direct proportion to effort, i.e., equity. Under the conditions likely to be prevailing in aid project settings, namely cultural diversity and material deprivation, considerations of equity may well be particularly germane, especially perhaps if they somehow threaten the integrity of one's sense of cultural identity (see also Fechner's Law, Chapter 5).

To the extent that a range of host workers felt united in (pay) adversity (David and Turner, 1996), they might also apply the principle of equity on a *group* rather than exclusively individual basis. In management, the concept of 'external equity' (or social equity between one's own organizational group and those performing the same job in another organization or market) is widely recognized (for example, Landy, 1989; Makin *et al.*, 1989; Muchinsky, 1997). Recognition has also been given in development studies (for example, Vallance, 1996). In Zambia, for instance, Machungwa and Schmitt (1983) found that demotivation (tardiness, absenteeism, restriction of effort, and quitting) revolved especially around resolving *inequity between (ethnic) groups*.

We have argued elsewhere that exposure to Western forms of (work) culture is more likely to invoke pluralism than assimilation of Western cultural values. One of the reasons for the past economic successes of the 'Tigers' of East Asia may have been their psychological and socio-cultural 'carrying capacity', to create patterns of modernization that combined so-called 'modern' values with traditional ones (Marsella and Choi, 1993). The notion of equity may thus be rendered more relevant by the contemporary globalization of capitalism, and rising individualism generally (Bond and Smith, 1996; Cox, 1995). Equity though does not originate from, or 'belong' to, contemporary or Western belief systems (see the Book of Matthew, Chapter 20, and Keats and Fu-Xi, 1996, for clear cases in point). By the same token, pluralism is not unique to non-Western societies (Kao and Ng, 1996). One of the original battlecries of European communism, for instance, was 'from each according to their ability, to each according to their need' (Marx and Engels, 1948). More to the point, however, what one person considers just may alternate between equity (when all that matters is the task), and need (in adversity), and egalitarianism when relationships are uppermost (Assmar and Rodriques, 1994; Leung and Park, 1986).

The overall point to be extracted from all of this may seem obvious, but it is often overlooked by (social) scientists possibly oriented to seek 'one truth' (Hofstede and Bond, 1988, p. 20). No one culture or context has a monopoly on the principles of equity, need, or equality. In addition, of course, equity-focused expatriates can believe in *need*! Now, perhaps, we can see more clearly how people can be motivated by more than one motive at a time, and especially by more than one 'cultural' value at a time, namely by motives that are *in concert*.

One particular case of psychological pluralism is evident at work. From Table 9.1 (see p. 152), Malawian hosts were more likely than their predominantly European and American expatriate counterparts (both on international *and* local salaries) to agree with propositions stating that local salaries were inequitable. This clearly suggests that the Malawians were applying some kind of additional value(s) to a criterion of fairness. That may have been the yardstick of preserving socio-cultural 'face' (see Chapter 4), and/or have been due to differences in perceptual threshold (see Chapter 5). In any event, these hosts at home evidently possessed a comparatively keen sense of social (in)equity.

Distorting input

Underpaid workers may sometimes be prevented from reducing their input by behavioral means. For example, their humanitarian commitment to the work, or sense of community loyalty, may be too high or too compelling. In this eventuality, the principle of equity predicts that they might also engage in various *cognitive* distortions, specifically reducing what they *think* of their own inputs. The counterintuitive (and contentious) prediction being derived here is that glaringly unjust and persistent underpayment may, eventually, begin to undermine one's self-confidence and respect. Let us reiterate, too, that this process would probably be a gradual and insidious one and, because of its possibly self-deceptive nature, not necessarily conscious.

155

One possible pathway toward such self-derogation would be the process of dissonance reduction (see Chapter 6). Lower salaries are somewhat less humiliating if you somehow deserve them, so the underpaid might begin to convince themselves of this. Dissonance reduction, however, may not be the only salient factor in those pluralistic (and relatively tolerant) cultural settings that often characterize 'developing' countries. There are at least two other theoretical social processes by which self-depreciation could occur.

One of these is the more emotionally neutral self-perception mechanism (see Chapter 6). There, one calmly observes oneself to be earning less, inductively reasoning that this may therefore be deserved. Such 'cold cognition' is more likely when one is relatively low on commitment (Bem, 1967, 1972), say if comparatively low salaries have already had time to cause some demotivation, via dissonance reduction.

Alternatively, dissonance is ruled out when one cannot perceive any *choice* in one's actions; say, when one is more obliged to work for money because one lives in a 'developing' country. In that event, smaller incentives tend to have smaller impacts on motivation to work (Linder *et al.*, 1967). Again, this could be relevant to comparatively underpaid development workers. In developing countries, that is, the economic necessity of work is, almost by definition, likely to be more forcefully apparent.

While there is evidently some uncertainty surrounding the precise theoretical process, or processes, by which hosts' input could become distorted, the evidence is unequivocal that the outcome will be distortion. Equity studies have found that disadvantaged groups *expecting* to be treated inequitably do tend toward privately *accepting* that treatment (Austin and Walster, 1974). In a prison-simulation study in the United States, Zimbardo (1982) found that students role-playing prisoners eventually developed a sense of worthlessness and docility toward their prison guard masters, which is possibly an example of how a deterioration in intergroup relations can rapidly escalate into social conflict when there is a power differential to begin with (Tajfel, 1978). In Australia (Marjoribanks and Jordan, 1986) and in New Zealand (Lynskey *et al.*, 1991), both Anglo-Europeans *and* indigenous peoples reportedly hold the indigenous groups in relatively low esteem.

In the area of human development, several studies have found that disadvantaged children who are old enough to have a sense of identity will prefer to play with a doll that is visibly from a majority rather than a minority ethnic group (see, for example, Augoustinos and Walker, 1995; Berry *et al.*, 1977; Bourhis *et al.*, 1973; Vaughan, 1978). In the area of health and welfare, a similar suggestion has been made about concentration-camp inmates, the terminally ill, and the profoundly depressed (Lerner and Miller, 1978).

Several socio-political commentators would agree with the observation that deprivation breeds self-deprecation. These include Franz Fanon (1985, 1991) in Algeria, Memmi (1966) in Tunisia, Rivera (1984) in Puerto Rico, Bejar Navarro (1986) in Mexico, Montero (1990) in Venezuela, Alatas (1978) regarding the archipelagos of South East Asia, and Malcolm X in the United States (Haley and X, 1987). The last-mentioned commentator, for instance, described how African Americans needed to cast off the stigma of being 'knee-grows' (i.e., short in psychological stature relative to the white man). These widely scattered and independent commentators, as well as social scientists, have converged on the conclusion that the repeated humiliations of life under colonial (and neocolonial) rule may sometimes, eventually, result in a devalued sense of socio-cultural identity, i.e., in an internalized inferiority complex.

Double demotivation

Overall, therefore, there is is a wide and varied body of theory and evidence to support the prediction that the differential between host and expatriate salaries may – surreptitiously and decrementally – result in a devalued sense of cultural identity, lowered aspiration, and demotivation. Ultimately, we believe, such demotivation is likely to impact on the aid project itself. When we consider also the arguments and evidence reviewed in Chapter 6, salary differences may be unintentionally but seriously undermining the viability of international aid, not only by demotivating expatriates, but host counterparts too. There would, in effect, be a *double demotivation* (Carr and MacLachlan, 1993; MacLachlan and Carr 1993a)! Perhaps this (partly) explains why in his review of 277 USAID projects in Africa, Rondinelli (1986) found high turnover rates of both donor and host-country counterparts in TC, 'double turnovers' that were very damaging to aid projects themselves.

Managing it

What will count as an acceptable salary differential is often an empirical issue (Argyle, 1989), and the literature suggests that aid agencies might invest in surveying both groups in the particular aid partnership, with a view to determining relative preferences for, and acceptability of, a pay system based on equity (for example, implementing productivity bonuses), equality (for example, drafting local contracts for all workers), or need (for example, taking into account socio-economic backgrounds). This could be done at the pre-implementation phase of an aid project (Psacharopoulos, 1995). Such considerations are basic to the widespread concept of Organizational Development (Srinivas, 1995), as well as addressing ongoing concerns about the need for greater economy and efficiency in aid-organization salary budgets (for example, Dawkins, 1993; Dore, 1994; SBS, 1996). Such concerns would naturally be shared by aid organizations and government departments in the host country (World Bank News, 1996).

Productivity bonuses have been shown to be effective in a number of work settings (for example, Florkowski and Schuster, 1992), including human service organizations (for example, Weisman *et al.*, 1993). Restructuring pay awards, scaling-down for expatriate contracts and scaling-up for local ones, has been seen as a viable option in Papua New Guinea (Richardson, 1987). In fact, many universities in developing countries, as well as major aid organizations like the United States Peace Corps, British Volunteer Service Overseas, and Australian Volunteers Abroad, already have a policy of paying salaries at or near the local level. The option to try and match background needs across guest and host workers is consistent with the widely used concept of 'fit' between employee and organization, a concept that has often been related to job satisfaction (Hesketh and Gardner, 1993).

Salary differences, too, will often remain an inescapable reality to which personnel must be 'fitted'. The most pragmatic, 'damage-control' approach here may be to select aid workers who are comparatively insensitive to inequity. It should be possible to probe for such tendencies during a selection process, either by structuring the interview and/or using appropriate selection tests (see Chapter 7). As we have seen, McLoughlin and Carr (1997) found that individuals who tested 'equity *sensitive*' were more likely to become demotivated by underpayment. That is, they were more ready to quit working on an otherwise stimulating task at the first available opportunity. The inequity angered and belittled them substantially more than it did others who were less 'equity sensitive'. Women, too, may be less concerned by salary gaps, particularly in all-women teams (Brockner and Adsit, 1986).

Envy (and jealousy)?

In the case of double demotivation, development workers from both sides of an aid partnership may become (de)motivated, despite their different cultural backgrounds. Interventions may not be quite so straightforward, however, when the donor and host fundamentally *dis*agree on what precise form of social justice should prevail. That is, hosts may find themselves being pulled and pushed in *differing* directions, say by a foreign system of working, on the one hand, versus a 'traditional' system of social conduct and moral obligations, on the other.

The resulting ambivalence may create what are called 'approach–avoidance' conflicts. As an example of how this might happen, consider that the culture of Western organizations is frequently imbued with an ethos of individual competition. For donors at home, this ethos may simply be a reflection of the capitalist values that are integral to 'organizations'. Understandably, the recent ideological 'defeat' of communism has brought a groundswell of support for this ideology of competition (Cox, 1995), a resurgence of faith in 'the free market', which has resonated within the international aid community.

In Eastern Europe and indeed generally, 'aid for trade', and the development and training of 'human' resources, have become watchwords of development (for example, AusAID, 1995; Booker-Weiner, 1995; Gertzel, 1994). At the same time, 'international assistance agencies have devoted a large portion of their financial, administrative, and technical resources to improving organizational and management capacities in LDCs' (Rondinelli, 1986, p. 421; see also, Thomas,

1996). If 'improving' means making more 'competitive' at all levels, the notion of individual competition is once more being exported via, and promulgated in the name of, international aid.

Yet there is a new paradox in all of this. The exportation of our own cultural values runs directly against the tide of growing recognition that development projects can become more sustainable by respecting, rather than attempting to override, cultural (and sub-cultural) traditions (for example, Adler, 1986; Blunt and Jones, 1992; Marsella and Choi, 1993; Srinivas, 1995). Exactly the same point has been made in relation to the fit between aid-project organizational culture and socio-cultural environment (for example, Kaul, 1989; Stone, 1992). One root of such problems, as we see it, may often be intercultural. In less individualistic host countries, an ethos of personal promotion may clash seriously with traditional societal values that place collective life above personal gratification. At the individual level, the employee may once again be 'caught up' between the demands of Mazrui's two moral systems, one stressing individual competition and the other group cooperation.

At the sociological level, as Kohnert (1996) has pointed out, modernization and the market economy have witnessed, in developing countries, an *in*crease in the incidence of magic and witchcraft. At the social-psychological level, the consequences of a self-oriented achievement motivation might therefore include envy (covetousness) from co-worker peers and jealousy (in the dictionary sense, jealousy equals guarding something) from superiors. These would represent social-organizational forces that we have likened to a motivational gravity (Carr and MacLachlan, 1997).

According to Trompenaars (1993), one of the principal differences between cultures of donor and host concerns achievement versus ascription, namely whether status is normally accorded via personal achievements or through social indicators such as age, education, or lineage (p. 29). We believe that this particular cultural watershed, related perhaps to what Hofstede (1980) called 'power distance' (see Chapter 4, this volume) has profound and wide implications for the provision of international aid. That is, demotivation may be unwittingly promoted by some aid projects and organizations because of their somewhat unquestioning promotion of 'achievement', which implies a certain disregard and even disrespect for cultural-traditional ascriptions and ascriptive norms. As such, aid itself may sometimes pave the way for feelings of social inequity and resulting conflict at work.

The fundamental problem with such a monocultural approach is that it may not be sufficiently mindful of the cultural value placed, in parallel, on ascription. It risks setting in motion an intercultural conflict in which any 'approach' toward personal achievement can easily be countered by 'avoidance' of potential social costs from colleagues and superiors. In cost-benefit terms, encouraging self-promotion rather than social responsibility may lead people to begin to become *wary* of personal success, so that they anticipate social forces designed to re-assert the status quo. In Malawi (Bowa and MacLachlan, 1994), as in Sweden

(Schneider, 1991), individuals may go so far as to refuse a promotion, on the grounds that it may bring more social problems than it is worth. Manifestly, such outcomes could conceivably impede organizational 'development', and we can easily envisage individualism and collectivism 'pushing' and 'pulling' in the same manner.

The anticipated counterforces to individual achievement may be centripetal (the opposite of centrifugal, i.e., pulling inward), representing efforts by co-worker peers to restore group cohesion, or vertical, designed to keep subordinate workers in their place in the organizational 'tree'. The tactics in question here will probably be familiar to many readers, and they include (from co-workers) backbiting, gossiping, and conspiring generally, while bosses may block promotion, steal ideas, or transfer individuals to another department. Job insecurity, and hence any tendency to protect one's position, may be amplified by the internationalization of market environments and globalization of 'free-market' competition (Kao and Ng, 1996).

The latter point reminds us again that tendencies to rein in (or, in a strict organizational sense, Pull Down) successful co-workers, and to Push Down on budding and ambitious subordinates, are not (nor have ever been) the sole property of so-called 'developing' countries (Schoeck, 1969). Pull and Push Down tendencies have, for instance, been documented in the United States (Mayo, 1949; Schoeck, 1969), Australia (Colling, 1992), Sweden (Schneider, 1991), and Hong Kong (Wong, 1996), as well as among insecure White managers in Zimbabwe (Kaplinsky, 1995; Posthuma, 1995) and within international aid agency headquarters (Harrod, 1974; Thomas, 1996).

Nonetheless, employees from collectivistic societies, in the long run, may be *relatively* more likely to apply their societal standards at the workplace (Templer, *et al.*, 1992). Insofar as collectivism entails centripetalism, it may also promote Pull Down. With regard to Push Down, there is likely to be a certain amount of power distance, as well as greater job insecurity if the economy is a *less* 'developed' one, competing in an ever-globalizing marketplace. To these extents, and in the long run, organizational systems in aid settings may contain proportionately more Pull Down (versus Push Up) from co-workers, as well as Push Down (versus Pull Up) from superiors.

The idea that there may be proportionately more net 'motivational gravity' in more collectivistic and developing countries is supported by a wide variety of evidence. In relation to Pull Down, for instance, successful Zambian, Malawian, Kenyan, and Nigerian employees may be seeking spells to protect themselves from the witchcraft of workers with whom they are competing (Blunt, 1983; Bowa and MacLachlan, 1994; Lawuyi, 1992). In those same countries, managers may also be taking actions (such as transferring and disempowering over-achieving subordinates) designed to protect their own insecure positions in the organizational hierarchy (Blunt and Jones, 1986; Jones, 1988, 1991; Kiggundu, 1986, 1991; Machungwa and Schmitt, 1983; Montgomery, 1987; Seddon, 1985). Similar patterns have also been reported from collectivistic and hierarchical societies located within the South Pacific region (for example, Traynor and Watts, 1992) and in East Asia (for example, Hui, 1990).

Expatriates, including those working for aid organizations, are not exempt from these tendencies toward Push Down. In the cross-cultural context of working in 'developing' countries, they may unwittingly demotivate, discourage, and disparage their local colleagues for failing to live up to their own expectations (Fox, 1994; MacLachlan, 1993b). Inasmuch as expatriates are often brought into senior positions, this will Push Down their local colleagues, something that was vividly illustrated in 'The case' (see Chapter 1). The same tendency has been discerned among expatriates writing about their 'experiences' with African organizations (Munro, 1996), while researchers based in Western universities sometimes exploit junior colleagues in developing countries by poaching raw data (Moghaddam, 1993). In Vanuatu, aid expatriates may jealously guard their managerial positions against indigenization programs (Traynor and Watts, 1992). Evidently, the ideal of the expatriate 'working oneself out of a job' can, on occasion, represent a conflict of interest with job security. The result may therefore be a form of Push Down that characterizes many international aid projects in the developing world.

The concepts of Pull Down and Push Down each represent the end of a bipolar continuum, with their antonyms being Push Up and Pull Up, respectively. These bipolarities enable us to envisage a matrix of four distinctive combinations, or *gravitational forcefields*, of subjective norms (see Chapter 2; Paul, 1995). This matrix forms what we have termed the 'Motivational Gravity Grid', which is presented in Figure 9.1.

The grid can be thought of as a taxonomy of organizational cultures, defined in terms of the social-psychological pushes and pulls, exerted on any one individual employee. Such cultures do affect business performance (Klitgaard, 1995). Our grid, though, is about the individual in relation to others in the organization, complementing earlier analyses in terms of individual and company differences in 'submission–dominance' and 'hostility–warmth' (see Buzzotta *et al.*, 1972). Motivational gravity is intercultural and dynamic in character.

In the brief review of evidence that now follows, the emphasis on *organizational* culture should remind the reader that there may be as much variation within countries as between them (Ralston *et al.*, 1992). Although we have predicted that organizations in developing countries are, on average, more likely to be categorized in the (pull and push) 'Down' quadrants of the grid, we are not insinuating that this will always be the case. We wish to avoid stereotyping organizational cultures in terms of national culture. Intergroup dynamics operate in all workplaces, and it would be foolish to presume otherwise.

Figure 9.1 The Motivational Gravity Grid
Note: The authors are grateful to Matthew Hodgson of Newcastle University, Australia, for
   illustrating their ideas so vividly.
Source: Carr and MacLachlan (1997).

Pull Down–Push Down

In Malawi, we have presented managers and trainee managers with scenarios
depicting individual achievement at work, and have asked them to predict what is
likely to happen next (Carr, MacLachlan, Zimba, and Bowa, 1995a). Almost
unanimously, these competitive scenarios, or mini-narratives, have resonated
with the respondents' own perceptions that the Malawian workplace is highly

competitive. There has also been a very definite and resounding expectation that such competitiveness will be countered by sharp Pull Down and Push Down reactions. These will include, for example, ostracism by peers and plagiarism of subordinates' ideas, as well as the use of witchcraft and witch doctors. These expectations were clearest of all among the more experienced managers, who emphatically agreed that one should not encourage others to perform better than oneself. As we had anticipated on theoretical grounds, respondents' explanations for their answers hinged not only on competition, but also around inverse resonance and job preservation. These findings converge with the most frequently cited causes of demotivation, tribal nepotism, and bad relations among workers and superiors in Machungwa and Schmitt's (1983) Zambian study.

Let us stress that we are *not* saying that competition is *necessarily* counterproductive. But what we *are* saying is that, *within* organizations, it *can* be. In the West, Senge argues that a self/ingroup-oriented focus in organizations often escalates into divisiveness, resulting in unnecessary and inefficient duplication and victim-blaming when organizational or departmental performance suffers (1992, p. 19; see also Cox, 1995).

The possible consequences of 'developing' an over-competitive ethos in organizations that operate within *collectivistic* societies is indicated by a study conducted at Malaŵi's major psychiatric facility (MacLachlan *et al.*, 1995b). There, some 40 percent of patients and accompanying custodians attributed the patient's admission to social-psychological factors resembling motivational gravity. This resemblance is nicely captured in the comment, 'He was bewitched by his workmates, because he works hard . . .' (p. 83). Although such bewitchment is often the *modus operandi* of motivational gravity in Malaŵi, it would be naïve and hypocritical to conclude that Westerners are exempt from magical thinking. The line between malice and witchcraft is often hard to draw, and as Schoeck reminds us, 'How often do we exclaim: "You *wanted* that to happen!"' (1969, p. 104, emphasis added).

Nonetheless, to the extent that human resources take on added importance in materially under-resourced settings, motivational gravity generally, and witchcraft in particular, may be particularly damaging to the development of less industrialized nations. Developing countries may be least able to afford an unbridled capitalism that simply provokes both Pull Down and Push Down reactions. If so, the issue of *managing* social vs. individual achievement may require considerably more ingenuity and thought than it has been accorded up to now.

Push Up–Pull Up

One way of using the grid to aid managerial decision-making would be to look for factors that discriminate Pull Down–Push Down organizations from Push Up–Pull Up ones, all else being reasonably equivalent. African and Asian organizational settings, for example, have been said to exert similar sorts of demands and constraints on their respective managers (for example, Blunt and Jones,

1992). Furthermore, there may be some common characteristics that tend to set effective managers apart in both 'developing' and 'developed' countries (Munandar, 1990), as well as in aid projects themselves (Munro, 1996).

A recent survey of Japanese managers indicated that organizations there may often be characterized by Push Up–Pull Up (Carr, 1994b). This is despite wider societal norms advocating 'The nail that stands out gets pounded down' (see Chapter 4; Markus and Kitayana, 1991, p. 224), plus some evidence that this popular adage is often adhered to in everyday life (Feather and McKee, 1993). Furthermore, Japanese society, like some developing countries', may also be relatively collectivistic and power distant (Hofstede, 1980).

Taken together, these findings indicate that motivational gravity is somehow being managed effectively in some Japanese organizations. It therefore becomes legitimate to ask whether there is any particular, i.e., discriminating, management practice among Japanese organizations. That is especially so, since Japanese management techniques appear to be spreading successfully into developing countries (for example, Kaplinksy, 1995), including African ones (for example, Posthuma, 1995).

One particular worker sentiment did in fact stand out, one that has been practiced in Japanese organizations for many years (Clarke, 1979; Hui, 1990; Kim, 1994). This was a feeling of interconnectedness; of a common fate. Elsewhere, in the management training literature, this feeling is known as 'positive interdependence' (Johnson and Johnson, 1990). If one worker does well, that is felt to reflect positively on everyone connected to him or her through their common employing organization, department, etc. (a corollary of this is that punishing anyone for their achievements would reflect negatively on the entire group). Positive interdependence may thus act as an effective buffer against the often bruising nature of competitive rivalry. Indeed, similar metaphors have been described and recommended in China (Volitional Emulation, Sheridan, 1976), Indonesia (Gotong Royang, or 'collective self-help', Harriss, 1995), and Malaŵi (Collective Ascension, Bowa and MacLachlan, 1994).

How can this feeling be developed? As we ourselves have argued, psychological studies have indicated that this crucial sense of team can often be inculcated by setting a series of mutually valued, or 'superordinate goals'. These are ends that require the participation of everyone (Sherif, 1956) without necessarily blocking out self-expression (and individual reward) completely (Aronson et al., 1978). They have recently been recommended by UNICEF as a major vehicle for fostering a sense of positive interdependence (Fountain, 1995), in recognition of the growing need to counter some of the dehumanizing, and desocializing effects of globalization (Cox, 1995).

Aronson et al.'s (1978) educational blueprint for fostering (interethnic) team spirit, and one that proved very successful, consisted of giving each student an integral task in the overall team problem – a method they named the 'jigsaw' technique. An alluring feature of this technique is that it gives vent to both collectivism and individualism (see Chapter 4). It recognizes that 'rewarding

individual achievement *versus* having everyone feel valued' is to fail to see the possibility of a win–win outcome (Senge, 1992, p. 66, emphasis added). A nice illustration of how this might actually translate into the workplace is given in Figure
9.2. This example comes from contemporary practice in Australia, and focuses on
respecting both social and individual motivation through flexible pay systems.

There are grounds for believing that such jigsaw techniques might be particularly effective in collectivistic contexts, namely via their emphasis on individuality
within a group project. In Indonesia, for instance, 'The way in which production
is organized encourages team work and does allow for individual initiatives to be
exercised' (Harriss, 1995, p. 127). A similar argument has been made to explain
the success of quality circles in 'collectivistic' Japan (Smith and Bond, 1993).
Team incentives have also worked before at workplaces in Scandinavia, where
there is a tradition of 'Royal Swedish Envy' (Schneider, 1991; Schoeck, 1969), as
well as in Tanzania (Blunt and Jones, 1992), and in India (Blunt, 1983).

Like other East Asian Tigers, the Japanese have very successfully utilized the
traditional concept of a united family as a workplace metaphor (Hofstede and
Bond, 1988; Kashima and Callan, 1994). For example, workers are encouraged
to think of their supervisors as father figures. Carr (1994b) has suggested that this
particular interlinking and overarching (superordinate) metaphor may have been
crucial in the apparently successful management of Motivational Gravity found in
his study, and has raised the suggestion that it may apply elsewhere. In India,
J. B. P. Sinha and D. Sinha (see D. Sinha, 1989a) have already advocated the
application of a traditional family metaphor to organizational leadership. In
Kenya, recruitment from family networks has already proved to be a successful
method of generating company loyalty (Blunt, 1983). In Malaŵi, as in other
countries in sub-Saharan Africa, the traditional family continues to command a
great deal of respect.

Percentage Increments in Pay

	Exceeds target	4–5	6–7	8–9
The TEAM (or COLLECTIVE)	Achieves target	2–3	4–5	6–7
	Misses target	0–1	2–3	4–5
		Misses target	Achieves target	Exceeds target

INDIVIDUAL

Figure 9.2 A performance pay matrix
Source: Adapted from Carbet and Chikarovski (1997).

To sum up, therefore, traditional family metaphors may have a widespread scope for aid human-resource-management training in general, and for managing motivational gravity in particular (but, see our discussion of gender and motivational gravity below).

Pull Down–Pull Up

Other cultural settings and social contexts may be better suited to different metaphors. Australians, for example, are often relatively egalitarian (Cox, 1995; Feather, 1994). Their rejection of authority has historical origins in a reaction against former colonial masters (Conway, 1971). People's lives often derive meaning and motive through socio-historical and cultural narrative or scripts (Kashima, 1997), with roles that may well include intercultural dynamics. Social anthropologists, for instance, are frequently (partly) attracted by the 'romantic aspects of fieldwork . . . The old stereotypes probably still motivate the expatriate' (Littlewood, 1985, p. 196). Meanwhile, and possibly as an extension of antipathy toward hierarchy and authority, Australian workers may sometimes find themselves caught between Pull Down from egalitarian co-worker peers and Pull Up from individual bosses (McLoughlin and Carr, 1994; Orpen, 1990, 1993). In Australia, the 'quiet achiever' has to be quiet (and loyal) before his peers and yet 'achieve' for his or her boss (a pluralism sometimes required also in social life, Cox, 1995).

Sport is a radically different domain, however (Karpin, 1995). In keeping with their historical rejection of and by authority, Australians extol its 'level playing field' (Colling, 1992). A sporting metaphor may therefore provide a culturally appropriate, alternative metaphor for managing motivational gravity in Australia (Chidgey and Carr, 1996). This is also conceivable among workers in a stratified (and sports-mad) society such as Brazil (Todaro, 1994, p. 134; World Bank, 1995b, p. 221), where students prefer to allocate educational resources according to a principle of egalitarianism (Rodriques and Assmar, 1988). This may translate to organizational settings too. Egalitarianism there has involved collapsing the conical management rings in Figure 4.1 (see p. 76) into a series of concentric circles in a single plane (Bedeian, 1995). This radical compression has met with much success (Semler, 1994).

Let us remember, though, that precisely which metaphor, or 'script', will succeed is ultimately both an empirical and a contextual matter (Mendenhall and Wiley, 1994). In Singapore, for instance, sport is widely regarded as a second choice to more 'intellectual' pursuits.

There is growing concern in Australia that a 'Tall Poppy Syndrome' (tall poppies are high achievers that must be chopped down by those below) may be smothering workplace innovation (Anderson and Alexander, 1995; Brewer, 1995; Karpin, 1995; Robbins et al., 1994). In response to that concern, recent research has investigated the applicability of 'group polarization' (Moscovici and Zavalloni, 1969), which was discussed in Chapter 3. The technique has been applied, for instance, to interethnic relations (Myers and Bishop, 1970), and attitudes toward fellow employees (Myers, 1975).

Group discussion has produced a net gain in positive attitudes toward achievement, by augmenting pre-discussion nominal tendencies toward encouragement rather than discouragement (Carr, Pearson, and Provost, 1996). It may be especially useful during the early (Smith and Bond, 1993), induction phases of employment (Baker, 1995), but may also come into play to some extent in more established quality circles. To the extent that group techniques are consistent with collectivist work values, they might therefore represent a promising vehicle for nipping motivational gravity 'in the bud', either during aid-funded training projects and aid organization induction programs, or when preparing local communities for the distribution of aid-project benefits.

Push Up–Push Down

Managers may also require training on how to provide equal opportunities in the workplace for women, a problem that appears to be fairly universal in scope (Adler, 1993). In Australia, for example, outstanding work by women may attract downgrading (i.e., Push Down) rather than support from men (Forgas, 1986), both in staff development (Ellerman and Smith, 1983) and in terms of selection (Smith and Carr, 1997). Similarly, in some developing Asian economies, women may sometimes experience difficulty in having their achievements recognized by male superiors (Daroesman and Daroesman, 1992; Kaur and Ward, 1992; Nasir and Ismail, 1994). Like motivational gravity generally, these wide-ranging findings suggest that there are often human-factor limits to the success of equal-opportunity projects and programs.

Another key issue here concerns the reactions of other women. The continuing strengths of the Women In Development (or WID) movement may mean that women will likely support other women, rendering them Push Up. The empirical evidence, however, is rather more equivocal (Ashkanasy, 1994). Although one of the hallmarks of female development may be intimacy with friends of the same gender (Schultz, 1991), women described as seeking 'masculine' occupations may be seen as undesirable friends, by both men *and* women (Pfost and Fiore, 1990). Influential theorists in social psychology (Festinger, 1954) and in sociology (Schoeck, 1969) have argued that we tend to compete against those we feel we resemble. Thus, women might conceivably be anticipated to exert some degree of Pull Down on their female colleagues.

From Australia, Power (1994) reports that women subordinates withhold support from women superiors. As these analyses continue, it might become possible for psychological research to further inform aid-sponsored management training projects, as well as international aid organizations themselves, on the issue of managing gender equity in the workplace. Using the grid in Figure 9.1 (see p. 162), it may prove possible to isolate practices that discriminate between organizations that manage to convey a sense of Push Up (versus Pull Down) with respect to women and development. The same would apply to Pull Up (versus Push Down) by (male and female) bosses.

Concluding Remarks

Whether we are dealing with acute salary differences between expatriates abroad and hosts at home, or with the social costs of individual versus social achievement, the sustainability of an aid intervention may depend to a significant degree on the intercultural dynamics of social equity. Insofar as these human factors operate at the psychological level, aid efforts will require, or at least stand to benefit from, learning to understand, and to manage, behavioral phenomena like double demotivation and motivational gravity. We have tried in this chapter to mark some small beginnings for that important, but so far sorely neglected task.

10

TOLERANCE AND DEVELOPMENT

Cassen (1994) has argued for the adoption of single-country case studies in order to understand the true impact of aid. He argues that there are too many political, economic, and social differences between developing countries for us confidently to draw conclusions about development initiatives across many countries. By focusing primarily on one country the interplay between different factors can more accurately be assessed and evaluated. The backbone of this chapter will therefore be a review of some of our own research on health in Malaŵi. However, we also wish to refer to research from elsewhere that resonates with our emerging themes of (1) pluralistic approaches to health, and (2) tolerance of radically different perspectives.

Most people, in most countries, have to make choices about from whom they should seek help when they are not feeling well. In Europe, for instance, a person who experiences backache may choose to go to a general medical practitioner, a physiotherapist, a faith healer, or somebody else, in the hope of alleviating their suffering. Each of these healers may offer quite different perspectives on treating an ailment, and their approaches will be accepted to differing degrees within the dominant culture. In addition, healers from minority cultural groups may offer 'alternative' forms of healing, which members of the majority cultural group can either accept or reject as a valid form of therapy. These 'foreign' therapies and therapists are rarely incorporated into the mainstream health services of the dominant cultural group. Rather, they remain marginalized and disempowered, probably because to some extent they are seen as being too different from what members of the majority cultural group have been brought up to expect. Consequently, financial and human resources are channeled into providing a health care service which reflects the dominant model of health within that society. In Western cultures, this is usually the biomedical model, which endorses a molecular and reductionist explanation for the experience of illness.

In many of the world's developing countries, however, the health services provided by governments do *not* reflect their society's dominant models of health and illness. Instead, they reflect 'foreign' models. These models have often flourished alongside, but independently of, extant local models. People are presented with a more obvious choice between healers and health-care systems. In many developing countries, it is not the minority belief system that is disempowered: it is the *majority* belief system! Foreign aid has aided the development of its own 'foreign' systems of health care, rather than supporting the development of indigenous systems. For those who have never been to a developing country, it must be difficult to imagine that international aid has generally supported the development of a system of health care which the majority of the population considers to be foreign. An analogous situation to this might be that of Malaysia giving 'aid' to Germany, but only for the development of acupuncture as a treatment. This, of course, would make very little difference to health services in Germany, because Germany has a much larger economy that Malaysia, and its citizens enjoy a relatively high standard and quality of life. However, in the case where the giver of aid is economically much more powerful than the receiver, and where the health status of citizens in the receiving country is relatively poor, the receiving country and its citizens are disempowered.

In many developing countries, therefore, the establishment and development of biomedically oriented health services can be seen as a form of cultural invasion – an attempt (albeit benevolent) to colonize the mind and the body, by caring for it in a foreign way. This is not to deny that developing countries have benefitted from foreign aid channeled through Western-oriented health services, or to suggest that one culture cannot benefit from incorporating the health practices of another. However, it is important to appreciate that most people in developing countries are presented with choices concerning their own health care, choices which are quite different from the sort of choices that the majority of people in Western countries must make. We now consider the interplay between biomedically rooted models of health promotion and indigenous models of health care. As will be seen, sometimes there are significant differences between them and failing to take account of these may prohibit the effectiveness of internationally aided campaigns to promote health.

Health promotion: the message from the West

The World Health Organization (WHO) and three other United Nations agencies (UNICEF, UNESCO, and UNFPA) have collaborated to produce 'Facts for Life' (Adamson, 1993), which identifies eleven key child-health messages. These messages concern timing births, safe motherhood, breastfeeding, child growth, immunization, diarrhea, coughs and colds, home hygiene, malaria, AIDS, and child development. It is claimed to be 'the most authoritative expression, in plain language, of what medical science now knows about practical, low-cost ways of protecting children's lives and health' (1993, p. ii). 'All for Health' (Williams, 1993), also produced by UNICEF, is a companion to 'Facts for Life' and focuses on how effectively to communicate the basic health-promotion knowledge contained in 'Facts for Life'. By and large, this knowledge is not technical, and is well within the grasp of even illiterate people, if communicated in an appropriate manner. In each case, however, it is the biomedical model which underlies the rationale of the messages to be communicated.

'Facts for Life' and 'All for Health' are excellent resource materials, which provide a useful framework to which local practitioners can refer for guidance on general principles. While it is acknowledged by the authors that 'giving the facts' is not enough, there is a danger that some people will assume a 'linear' model of health promotion. In such a model it is assumed that the stronger the message, and the more frequently it is communicated, then the greater will be the desired behavior change. However, sometimes community norms may constitute anti-health-promotion ideas (MacLachlan, 1996c) and these ideas may oppose the adoption of health-promotion messages, such as those in 'Facts for Life', even when those messages are optimally communicated.

Many industrialized countries have witnessed the loss of what Sarson (1974) refers to as 'The Psychological Sense of Community', a loss which he has dubbed 'the central problem of social living' (p. viii). Ironically, the strong psychological sense of community apparent in many developing countries, especially in rural areas, may be one of the strongest barriers to the adoption of more 'healthy behaviour'. We shall now consider the community contexts in which the messages in 'Facts for Life' must take root and contrast them with 'anti-health-promotion' ideas, which will be seen to have an important community function in their own right. The 'data' concerning health beliefs, presented below, come from a variety of sources (MacLachlan, 1994). However, the major source was critiques of 'Facts for Life' messages, written by third- and fourth-year psychology students at the University of Malaŵi. These students (N = 38) worked in groups to identify indigenous beliefs which might impede the effectiveness of the 'Facts for Life' messages. Each group produced a report and these were content analyzed. While we make no claim for the representativeness of the data, it is worth mentioning that, owing to the quota system of admission to the University of Malaŵi, a roughly equal proportion of students is admitted from each geographical region. Consequently, the majority of students in this sample originated from rural communities. Three-quarters of students were male, reflecting the gender bias across the university generally. We are indebted to these students for their reports. Although eleven health issues are highlighted in 'Facts for Life', for the sake of brevity, we review only two here, diarrhea and immunization.

Diarrhea

Diarrhea is mainly caused by poor hygiene and a lack of clean drinking water. By producing dehydration, it kills millions of children every year and contributes to malnutrition. Table 10.1 gives the seven 'Prime Messages' in 'Facts for Life' regarding diarrhea. We now consider some community normative beliefs about diarrhea that may prohibit the effectiveness of promoting healthier behavior.

Developmental stage Rather than being seen as a sign of danger or illness,

Table 10.1 The 'Prime Messages' for promoting healthy behavior regarding diarrhea

---

1. Diarrhea can kill children by draining too much liquid from the body. So it is essential to give a child with diarrhea plenty of liquids to drink.
2. When a breastfed child has diarrhea, it is important to continue breastfeeding.
3. A child with diarrhea needs food.
4. Trained help is needed if diarrhea is more serious than usual.
5. A child who is recovering from diarrhea needs an extra meal every day for at least one week.
6. Medicines should not be used for diarrhea, except on medical advice.
7. Diarrhea can be prevented by breastfeeding, by immunizing all children against measles, by using latrines, by keeping food and water clean, and by washing hands before touching food.

Source: Adamson (1993).

diarrhea can be seen as part of the normal development of the child. Logical support for this notion comes from the fact that virtually all children experience diarrhea, and that this is followed by the developmental milestone of teething. Thus, diarrhea is seen by some communities as a necessary stage of development, like teething, which must be successfully negotiated by the child. Because it is seen as 'natural', it is also understood to be harmless.

Cleansing mechanism In adults, diarrhea may be understood not as being caused by 'germs', but as the result of a self-triggered and self-stopping mechanism which the body uses to cleanse itself. On occasion, people may even take certain herbs to facilitate the process. As such, diarrhea is understood to be of benefit and also, possibly, to have a preventive function. Should the diarrhea persist, then intervention involves withdrawing water and other fluids, for fear that they will liquify food which has been ingested.

Sign of infidelity 'Tsempho' is the term given to diarrhea when it is understood to result from (i.e. to be a sign of) a man's having extra-marital affairs. If the 'victim' (the one who experiences the diarrhea) is the man – his wife is understood to be having extra-marital affairs – then he is also characterized by malnourishment and persistent coughing. Intervention in this case often constitutes a form of marriage counseling by elders in the community. If the diarrhea is epidemic, its cause is believed to lie in a member of the community deviating from social norms or from moral standards, perhaps again through infidelity.

Symptom of AIDS While not a traditional belief, community norms have become quickly established regarding the ways of telling if somebody has AIDS. With 12 percent of the total Malawian population at the time estimated to be HIV positive (Liomba, 1994) the survival value of identifying those with AIDS is evident. AIDS is also referred to as 'thin', a state which is understood to be brought about by diarrhea. Over-generalization from the symptoms of AIDS to the syndrome of AIDS is, ironically, often seen in educated people who have learned the signs of AIDS through public health-promotion campaigns.

These four community normative beliefs clearly incorporate both traditional and contemporary understandings of diarrhea. Each can be seen as a point of resistance for the adoption of the health promotion messages in Table 10.1. The developmental stage and cleansing mechanism explanations do not in any way identify diarrhea as a problem. The infidelity explanation sees it as a symptom of a problem. Quite appropriately, treatment focuses on solving the problem rather than on simply removing the symptom. The attribution of diarrhea to AIDS is a 'no-hope' message. In many developing countries, people are bombarded with the message that AIDS is incurable and deadly. Therefore, believing that you may have AIDS is unlikely to compel you toward taking measures to stop, or to prevent, further diarrhea. Other beliefs endorse the common-sense notion that somebody with diarrhea (producing watery feces) should not be given more water, as this would only worsen the problem.

Immunization

'Facts for Life' recommends immunization against tuberculosis, polio, diphtheria, whooping cough, tetanus, and measles. Although deaths from tuberculosis in Malaẁi almost tripled between 1985 and 1991, this increase is almost certainly due to the rising incidence of HIV/AIDS, and does not reflect the lack of success of the Malaẁian immunization program. For instance, coverage for measles immunization rose from 52 percent to 80 percent between 1985 and 1990 (United Nations, 1993). Nonetheless, measles remains a significant cause of mortality in young children (United Nations, 1993), and various community norms continue to present resistance to the immunization program. Table 10.2 gives the 'Prime Messages' regarding immunization from 'Facts for Life'.

Children should have measles A prevalent belief in many communities is that every child *should* have measles. The rationale is that if they don't, then the measles will be retained in the body and will eventually express themselves later in

Table 10.2 The 'Prime Messages' for promoting healthy behavior regarding immunization

---
1. Immunization protects against several dangerous diseases. A child who is not immunized is more likely to become undernourished, to become disabled, and to die.
2. Immunization is urgent. All immunization should be completed in the first year of the child's life.
3. It is safe to immunize a sick child.
4. Every woman between the ages of 15 and 44 should be fully immunized against tetanus.

Source: Adamson (1993).

life, perhaps with more severe consequences. Incorporated into the notion that having measles is not necessarily a bad thing is the fact that most contemporary mothers were never taken for immunization by their own mothers. In other words, 'If my mother didn't do it, then why should I?'

Distresses the children　The short-term effects of immunization can discourage mothers from having their child given prophylactic treatment. Firstly, there is the immediate pain of the injections and, secondly, the child may subsequently develop a fever in response to the vaccination. Socially marketing the preventive value of immunization can be complex, because the notion of introducing 'a little' of the illness into a healthy child's body 'to do it some good' is a curious one. From a longer-term perspective, some children who have been immunized *do* subsequently develop the illness they are supposed to be protected against (often children do not complete the full course of immunization), thus questioning the effectiveness of immunization as a preventive measure.

Mother's time burden　In rural Malawian communities, women have the primary responsibilities for taking care of the family's house, growing food in the 'garden', collecting fuel wood and water, and (where possible) running a small business in order to increase the family's income. These tasks require much time and effort. Often the clinics, where most immunization is undertaken, are many miles away with no motorized transport available. Simply to find the time to take a healthy child to the clinic, with the cost of neglecting other demands, can be difficult. From the mother's perspective, why should she (being overworked as it is) carry a healthy child on her back many miles to a clinic where the child will suffer?

Communication problems　Hospitals and clinics are often feared by rural communities. They are seen as 'places where people go to die'. They are also places where the least literate and most traditional people come into direct contact with some of the most literate and most 'modern' people, doctors and nurses. Often the gap is not only one of finding the right words, but also one of empathizing with a different perspective. Rural people may experience being 'talked down to' by health professionals. Some health professionals may also believe that they should not have to justify themselves to 'simple peasant people', and that directing patients authoritatively should be sufficient.

Immunization is the contemporary 'spell of protection', and the diseases it protects against have encountered alternative understandings, for instance measles, some of which may resist health-promotion efforts. The idea that children should have measles in order to 'get them out' has an almost cathartic quality to it. From the perspective of it 'distresses the children' a behavioral analysis of the short-term 'punishment' effects of immunization is evident. Mother's time burden and communication problems exemplify the difficulties in orienting to a new perspective, while still being 'locked into' traditional life styles.

Learning from the community

'Programs which seek to alter health practices and attitudes constitute efforts to change the local culture . . . [where these programs do not work] . . . it may well be that the attempted changes challenge established beliefs or practices which are more fundamental to the stability of the particular social or cultural system than is evident at first inspection' (Paul, 1955, p. 474). Many communities in developing countries have *developed* social systems which have been highly successful in maintaining social stability and harmony over many generations.

People are often most comfortable with what they are familiar with, and with what they know works. Here we do not mean 'works' for the better or for the worse; 'works' in this sense is not meant to be an evaluative term but more of a descriptive term. A social system can be seen as 'working' if it provides the members of that system with an understanding of the events and relationships within it. Thus, a social system which incorporates the notion of bewitchment can work well if it helps the people within the system to explain certain events, like the onset of malaria, or a person's decline into poverty.

Generally, health promotion efforts in developing countries are based on a Western rationale of reiterating a simple, clear, crisp message, designed to induce an intention for pro-health behavior. However, often these messages fail to acknowledge the importance of *negotiating* behavior change within the cultural context of traditional, and often materially impoverished, communities. For instance, McAuliffe (1996) found that Malaŵian women often labeled their own behavior, and the behavior of other women, as high-risk, with respect to contracting HIV. However, many of them, particularly those in polygamous marriages, had difficulty making the commitment to change. Some of these women were involved in high-risk behavior in order to meet basic subsistence needs. As such, it may be difficult for them to see the benefits of risk-reduction behaviors outweighing their costs. They may correctly see the benefits of risk-reduction behavior as long-term survival, while at the same time, see the cost of these behaviors jeopardizing their (and their children's) short-term survival (McAuliffe, 1996).

The 'negotiation' of behaviour change needs to take place at many levels. In the case of health promotion, this should involve the nature of the message to be promoted, source of the message, its medium of publicity, and so on. Collins Airhihenbuwa (1995) has argued that health-promotion practices in Africa are fundamentally Eurocentric, reflecting Western concerns with particular indices of health and simultaneously misrepresenting and devaluing local understandings of health. Airhihenbuwa argues for culture to be placed at the core of public health. Such 'cultural empowerment' examines micro (individual, family, community) and macro (national and international) influences on health beliefs and practices. One clear implication of this view is that the way in which aid workers from abroad may see a problem is not necessarily the way local people see it, if they see it as a problem at all.

175

Chamba use and misuse

A good example of this is the rather simplistic 'one way' in which most Westerners understand marijuana usage, in contrast with the multiplicity of ways in which it is seen and used in many developing countries. In a series of studies we have investigated chamba (marijuana), which is often used as an intoxicant in Malaŵi (Peltzer, 1989). Chamba related admissions to Zomba Mental Hospital, Malaŵi's major in-patient psychiatric facility, have been on the increase (Carr, Ager et al., 1994). However, chamba has had an important role in traditional Malaŵian ceremonies over many generations, and so the question arose of whether traditional uses of chamba or other more 'modern' factors, were linked to contemporary misuses of it. Let us emphasize that our interest was not in casual chamba use, but rather in the extent to which chamba use could be problematic for some individuals, that is, associated with their admission to a psychiatric hospital.

In the first detailed study of chamba use in Malaŵi, Carr and colleagues studied fifty chamba users who had been admitted to Zomba Mental Hospital because of psychiatric problems, comparing them to a control group of fifty non-chamba-using patients, admitted to the same hospital, and matched for gender, age, and admission date. The typical chamba-abusing patient was around 27 years of age, male, a subsistence farmer, taking the drug because it was the cheapest form of intoxication, reported 'seeing things clearly' (immediate effect) and general apathy (long-term effect), and, compared with other patients, was more likely to originate from a chamba growing area, less likely to have been raised by his natural parents (a slight effect), and had had more schooling. As well as providing useful clinical and epidemiological data, these findings began to suggest that a complex range of factors may be involved in chamba admissions in Malaŵi.

There are in fact a number of quite distinct ways in which chamba is used in everyday life in Malaŵi (Ali et al., 1994). These include occupational, medicinal, spiritual, social, and recreational functions. An example of its occupational use is where subsistence farmers take chamba to endow them with more energy, and improve their concentration, so that they may work longer hours in their fields. Chamba's medicinal functions include it being taken to help overcome social anxiety, being taken as a general tonic, and being used as a panacea for a series of serious physical ailments. Spiritually, chamba may be used to ward off evil spirits (as when it is left burning in a doorway at night). In the socio-cultural context, chamba has been used for many years as an integral part of ceremonies conferring rites of passage. Finally, its recreational use refers to taking chamba to relax or get a 'high'. Thus, the use of this drug, which (probably because of Western influences) is officially illegal, is much more complex in Malaŵi than in most Western countries, where its use is generally restricted to its recreational function.

The social aspects, triggers and effects of chamba use among psychiatric in-patients have also been investigated (MacLachlan, Page, Robinson, Nyirenda, and Ali, 1998). The vast majority of these patients did indeed identify chamba use as a serious problem and described a range of distressing consequences of its use, these included sickness, familial discord, and domestic violence. However, the reasons given for first using chamba did not include all of the uses described above. Some reasons given for initially using chamba were to prepare for work, to give courage before addressing an audience, to alleviate unpleasant thoughts and feelings, to ward off hunger, and to increase mental activity. While chamba use in Malaŵi appears to have transcended its traditional uses, traditional uses do not appear to be implicated in the initiation of the 'misuse' of chamba. Although some of the patients certainly had early experiences of chamba through traditional settings, the reasons cited for beginning to use chamba were primarily occupational, medicinal, and recreational, rather than the more traditional spiritual and ceremonial uses.

Given the varied uses and abuses of chamba in Malaŵi, how should the problems resulting from its abuse be dealt with? The unrestricted legalization of chamba in Malaŵi may be problematic, because recent research suggests that heavy use of marijuana may be a stressor for psychotic relapse, exacerbate schizophrenia and related psychotic disorders, and may even be a premorbid precipitant of psychotic conditions (Linszen et al., 1994). However, a blanket ban on its use (if this were possible, which we do not for one moment suppose it would be) would also prohibit the potential benefit which can derive from its use in spiritual, ceremonial, and medicinal contexts.

Overall, the case of chamba use in Malaŵi illustrates that even though Western countries may encounter similar health problems (marijuana abuse), the solutions to these problems must take the context of their presentation into account. The social meaning and function of marijuana in Malaŵi is quite different from its social meaning and function in the USA, for instance. Problems which have different meanings and functions require different solutions. It may be quite misleading for health professionals to assume that they understand a problem simply because they have encountered that problem in their own country.

Beyond this point, the case of chamba use in Malaŵi also illustrates pluralism and tolerance of pluralism. For example, many Malaŵians accept that there is more than one use of chamba, and that its use crosses many domains of life. Aid workers, through their lack of familiarity with chamba, are more likely to view it in one way, that is, as they view it in their own country. They are more likely to believe in just 'one truth' (Hofstede and Bond, 1988), the idea that different explanations are competing explanations and, therefore, that only one explanation can be adhered to. Breaking out of the 'one-world, one-truth' mind set is not easy, but in many developing countries there have been attempts, to integrate fundamentally different approaches to healing. We believe that this is a challenge that international aid also needs to address more directly.

Integrating traditional and biomedical healing

Hyma and Ramesh (1994) have recently reviewed the prospects for integrating traditional and Western forms of medicine in developing countries. They note that full integration of these approaches has not occurred anywhere, but that different models of partial integration exist. A key issue which they identify is whether we should strive for a truly integrated system, or for dual autonomous health-care systems. The Indian government, for example, provides financial and institutional support for the development of the traditional Ayurveda, Unani, and Siddha systems of medicine. The Sri Lankan government has gone even further, establishing a separate Ministry of Traditional Medicine. During the 1980s, various aid projects were established in Africa to explore the relationship between traditional and biomedical health services. Zimbabwe, being one of the most progressive countries in this regard, established joint seminars, and created a registry of traditional practitioners along with forms of professional self-regulation.

China has perhaps gone further than any other country in integrating traditional and biomedical approaches to health. In China, traditional medicine has a long history, with its own disciplinary controls, colleges, and research institutes. Every Chinese school for 'Western' medicine contains a department of traditional medicine, and every school for Chinese traditional medicine contains a department of biomedicine. Practitioners of both persuasions may be employed in modern hospitals, working together in community health centres. China is an impressive exception to the rule of biomedical dominance. Its approach to integrated medicine is reported to involve 5,000 medical doctors, who have undergone two years of additional training in traditional medicine. By contrast, in India and most other developing countries, the formal health system is clearly dominated by the biomedical model, and therefore geared toward an urban, modernized middle class (Hyma and Ramesh, 1994). The consequence of this is that traditional healers are cast as inferior paramedical and para-professional staff.

Hyma and Ramesh (1994) emphasize the psychological difficulties which health personnel experience in trying to integrate systems of healing which have 'different philosophies, theories, histories and geographies, different aetiologies of disease, educational and training backgrounds, and diagnostic and treatment methods' (p. 76). For example, Paranjpe (1994) suggests that the indigenous health system of Ayurveda and the Western biomedical approach are fundamentally incompatible. Epilepsy is described as apasmara in the Ayurveda system. However the diagnostic and therapeutic system of Ayurveda is based primarily on the cosmology and ontology of the Vaisesika and Sankhya-Yoga, emphasizing the five elements of nature (earth, air, fire, water, and space) and three bodily humors (Kapha, Pitta, and Vata). Paranjpe argues that such conceptual categories are 'fundamentally incommensurable with the conceptual categories of [the] modern biomedical system' (1994, p. 10). Such incompatibility is seen as prohibiting the integration of Ayurvedic and biomedical systems of medicine.

The difficulty of integrating such different systems may be especially daunting to those who have invested many years of study and professional training in one particular approach. Indeed, in some ways, it may be easier for people with less of a vested interest in one approach or the other to accept a plurality of approaches. We, and others before us, have tried to look at this phenomenon from the perspective of the consumer of health services, rather than the professional providers of health. Our own research has focused on the difficulty of psychologically integrating different models of health and illness, namely on what is often implicitly asked of 'developees' by health aid projects.

Cognitive integration: from dissonance to tolerance

'Cognitive dissonance' is an aversive state of knowing of inconsistency (see Chapter 6). Festinger (1957) suggested that the existence of this state triggered a 'dissonance reduction' process in which people seek to restore consistency in their belief system by distorting one or more of their beliefs. To apply this to traditional and biomedical health systems, if health personnel feel that these two systems are in some ways inconsistent or contradictory, cognitive dissonance theory suggests that either one should be rejected in support of the other, that certain beliefs about one or both of the systems should be distorted in order to accommodate the other, or that new beliefs are developed which help to restore the compatibility of the two systems.

While Festinger appears to have been aware of the possible cultural limitation of his theory of cognitive dissonance, it has, like many psychological phenomena, been implicitly assumed to be universally applicable. We believe that research on health beliefs in Africa and elsewhere illustrates the cultural limitation of the theory as originally stated by Festinger. That traditional and biomedical beliefs co-exist within the same individual is well known (see, for example, Barbichon, 1968; Jahoda, 1970, Peltzer, 1989). However, this co-existence has often been seen as somewhat problematic for health-service delivery, that is, simply reflecting a transient (and somewhat irksome) stage in the development process, often with the implicit assumption that one theory (the biomedical theory) will predominate as people become 'more developed'. Our research has sought to move things forward from this 'transitional' perspective, and to look on the ability to entertain more than one model of health as a contemporary resource for health-service delivery. To try and complement the existing research in other disciplines, and to add to the accumulating literature, we have emphasized the importance of empirically *quantifying* the cognitive interplay which occurs in the consumers of health services in Malaŵi (Carr and MacLachlan, 1996b; MacLachlan and Carr, 1994a).

We have also used the qualitative techniques more commonly found in other disciplines. In one study we surveyed a rural quota sample of Malaŵian males and females ranging in age from 5 to over 50 (Ager et al., 1996). Using a structured interview format, which allowed for open-ended answers, we questioned people about their beliefs regarding the cause, risk reduction, and treatment of malaria and schistosomiasis, and subsequently content analyzed their responses. Statistical analysis indicated that neither understandings of the cause of malaria or schistosomiasis, nor beliefs regarding prevention, were tied to preferences for treatment. The majority of individuals sought medical treatment for malaria and schistosomiasis despite the fact that many of them attributed the cause of these illnesses and risk of infection by them to non-medical factors. Traditional beliefs about the cause of malaria and schistosomiasis included non-material factors such as spirits and witchcraft.

An example of a traditional witchcraft explanation for malaria is where the 'victim' is believed to have been bewitched by friends and colleagues who were jealous of their success (see Chapter 9). This cause is seen as mediated through the social matrix, where some form of retribution (see also Chapter 11) is the force behind the victim's malaria. A traditional form of healing for this might involve the bewitching spell being removed. This treatment would be entirely consistent with the belief that malaria is caused by another person placing a spell. However, a belief in a biomedical treatment introduces another causal model. Thus belief in a traditional (witchcraft) cause and a contemporary (biomedical) treatment, each acting through different modalities, were found to co-exist in the illness model which some people held for malaria. This 'mixed-modal model' was also found in our previous studies of epilepsy (Shaba et al., 1993) and mental disorder (Pangani et al., 1993) which used an alternative, more quantitative methodology of Likert-type scaling, on which respondents rated their degree of belief in alternative causes and treatments.

Traditional attributions continue to be the most common form of explanation given by patients for their admission to Zomba Mental Hospital, Malaŵi's major psychiatric facility (MacLachlan et al., 1995b) and even in the relatively Western-ized setting of a private Catholic Girls School, traditional explanations were prevalent among those given to explain the cause of an epidemic psychological disturbance involving over 100 girls (MacLachlan et al., 1995c). Traditional explanations do not seem to be fading away. No doubt this is because they serve important social functions, as we noted at the beginning of this chapter. Tertiary students who are highly educated in the Western-styled education system still retain strong traditional beliefs. As long ago as 1970 Jahoda reported that, in Ghana, university students at the beginning of their tertiary education tended to reject the influence of traditional forces only to endorse them more strongly by the time they were completing their degrees.

To juxtapose the existence of mixed-modal models with Festinger's idea of cognitive dissonance, we have described an individual's or a community's ability to entertain more than a one-modal model, as *cognitive tolerance* (see MacLachlan and Carr, 1994a, for a review). We do not see this cognitive tolerance as a problem for, or side-effect of, 'development'. Rather, it is a significant ability or talent, which can be a resource incorporated into health care. Aid projects, like foreigners, depend on the tolerance of their hosts.

Others too have seen tolerance in this positive light. For example, Elliot *et al.* (1992) have described the ability of experienced (but not newly trained) Zimbabwian nurses to accept both traditional and contemporary biomedical models of suicidal behaviour as 'the emotional maturity to use comfortably elements of both belief systems' (p. 278). Perhaps this does not reflect only emotional maturity but cognitive maturity too, and a better fit with the psychology of their clients. This tolerance of more than one system of cause–effect relationships may reach beyond the 'one-truth' perspective so frequently, and intolerantly, endorsed by the West. Eastern philosophies also recognize the ability to tolerate inconsistency as a sign of maturity (Carr, Munro and Bishop, 1996).

There need be nothing illogical about endorsing various different belief systems. Nor is such pluralism restricted to developing countries. In many societies, people will pray for a (spiritual) cure for an ailment which they believe is biologically caused. However, the reason we have emphasized *cognitive* tolerance, especially in the Tropics, is that health-service developers have frequently not incorporated the explanatory models of health and illness extant in indigenous cultures which they have sought to develop. They have, instead, transplanted health services which are minority endorsed, foreign, and biomedically oriented. Such assistance has, either explicitly or implicitly, been *intolerant* of alternatives. Thus we return to the point that while cognitive tolerance for pluralistic health systems is probably present to some degree in all cultures, the degree of tolerance found in developing countries may, fortunately, be much greater than in those where the majority model of health is also the dominant model. The consumers of health services in many developing countries may therefore be much more tolerant of different approaches than are the Western-trained health professionals 'serving' them. If we consider health services from the consumer's perspective, rather than from the clinician's perspective, then the problem of 'cognitive integration' may be recast as a problem of 'cognitive *in*tolerance'. Our own research suggests that the consumer is more tolerant than the clinician.

Faith in treatment

To the extent that treatments are something 'done to' people, then tolerance of different treatments, could be seen as passive acceptance of the alternatives on offer. However, an experiment conducted by Zimba and Buggie (1993) nicely illustrated that tolerance for different treatments is an active process of engaging an individual's faith in those treatments. Zimba and Buggie investigated the placebo effect in relation to both a traditional Malawian herbal concoction and what appeared to be a commercial Western-styled medicine, each of which was pharmacologically inert. Subjects were falsely told that they could expect these preparations to influence their body. In each case, oral temperature (but not pulse rate) significantly increased following ingestion of the placebo. The two types of preparation were equally effective in producing a placebo effect, possibly within the same individual.

This experimental design shows that the placebo effect is present in the use of modern and traditional medicines. The placebo effect is, of course, a reflection of one's faith in a particular form of treatment. The placebo effect applies not only to medications, but also to the practitioners who prescribe them. MacLachlan (1997) has developed 'faith matrices' to describe the interaction of the placebo effect between faith in a treatment and faith in the practitioner. The effectiveness of both traditional medicine and biomedicine is bolstered by their clientele's faith in the preparations and practitioners. It is often, incorrectly and patronizingly, assumed that while traditional healers and their medicines may (sometimes) work through a placebo effect, biomedicine works through its 'legitimate' effect. Contrarywise, the effectiveness of any system of healing will be constrained by the beliefs which its recipients hold regarding its effectiveness.

Tolerance and the prevention of AIDS

In a further study of social-science students we explored the credibility ratings of different sources of information for the treatment and the prevention of AIDS. The students rated the credibility of Western-trained doctors and nurses, family members, traditional healers, radio and newspaper advertisements and others. Western-trained doctors and nurses were rated by far the most credible sources in each case. However, in the case of prevention, another important relationship emerged. Using the technique of factor analysis we found that ratings for these sources grouped together in a particular pattern. Doctors and nurses formed one group (or factor), government media formed another, and traditional healers fell into a third group, which we interpreted as reflecting community values. Furthermore, respondents' ratings of the credibility of traditional healers and the biomedical personnel (doctors and nurses) were not (cor)related. There was no relationship between how credible an individual rated biomedical clinicians to be, on the one hand, and how credible they rated traditional healers to be, on the other (MacLachlan and Carr, 1994b). Again, this is evidence for a degree of pluralism; a strong belief in one approach did not preclude a strong belief in another approach, at least for some respondents.

We have interpreted this finding to suggest that different forms of healing may have credibility in different settings. In rural communities, for instance, traditional healers may have greater credibility as a source of information about the prevention of AIDS, while in an urban hospital medical doctors and nurses have greater credibility. By working together it may also be possible to raise the credibility of each group in those settings where they are seen as less credible. Given the comprehensive network of traditional healers which exists in many developing countries (see Good, 1987; MacLachlan, 1993c) our results suggest that traditional healers may well have a role to play in the prevention of a deadly *contemporary* disease. It is worth pointing out that the results from the above study may, however, be an underestimate of the role of traditional healers, as our sample was taken from students highly educated in a Western-oriented education system. A similar exploration of credibility with a rural and less (Western) educated sample might reasonably be expected to give greater credence to traditional healers.

Recognizing complexity

Recently, we have taken this data one step further (Carr, Watters, and MacLachlan, 1996) by investigating the relationship between endorsing the credibility of biomedical practitioners and endorsing the relatively modern means of communication used by the government. Using non-linear regression techniques we found that for those people who did *not* endorse the credibility of community sources (such as traditional healers), there was a very strong linear relationship between endorsing the credibility of modern media and endorsing the credibility of biomedically oriented clinicians (doctors and nurses) as a source of information for the prevention of AIDS. However, for those people who strongly endorsed community sources (traditional healers, family members, etc.) the relationship which emerged between their endorsing the credibility of 'modernity' and 'clinicians' was not straightforward: there was only a very weak linear relationship, and evidence of a non-linear moderating function, known mathematically as a catastrophe (see Watters *et al.*, 1996, for a review).

What we are suggesting is that, in practice, for those who strongly endorse the credibility of traditional community sources, there is no straightforward linear relationship between endorsing the credibility of modernity and clinicians. A certain degree of belief in modernity may be associated, across the sample, with several different degrees of belief in biomedically oriented clinicians. Endorsing the credibility of modern sources of communication to a certain extent may coincide with endorsing the credibility of clinicians at more than one level, as having very low *and* very high credibility, *and* somewhere in-between (for example, via varying degrees of inverse resonance vs. resonance [see Chapter 6] between patients and clinician).

The most important point to be made is that although the relationship is non-linear, it is also statistically predictable. The equation which we derived to explain the variation in our sample's responses was able to do so to a statistically significant extent. This is perhaps the first evidence of being able accurately to predict apparently inconsistent or parallel health beliefs. The results, once again, challenge the 'one-truth' perspective of Western-trained clinicians and suggest that they should allow for health-service provision which encompasses different models of health.

We wish to reiterate that the need to develop such services is not limited to so-called 'developing' countries. Bishop (1996) has made a similar argument for Singaporeans, who have access to traditional Chinese, Malay, and Indian forms of healing, as well as biomedical 'Western' healing. Singaporeans commonly seek help from more than one of these systems simultaneously. Furthermore, compared with North American Caucasian students, Singaporian students use a greater number of cognitive dimensions to understand disease. They are also better able to synthesize 'Eastern' and 'Western' conceptions of disease. Perhaps, then, cognitive tolerance has less to do with developing countries, as such, and much to do with developing a more appropriate model of healing than a non-Eurocentric, non-'one-truth' model of biomedical reductionism. If this is the case then cognitive tolerance constitutes a considerable challenge to many aid agencies, which often have the origin for their programs and projects in Europe or North America.

Conclusion

Cognitive tolerance is a resource which could be built upon to provide a geographically and cognitively more comprehensive health service to people in many developing countries. While people in all countries are presented with health choices related to different ways of understanding their well-being, in developing countries the model of health endorsed by the majority of people is often not the approach to health care which government and international agencies have been trying to develop. However, the health issues to be confronted in developing countries require a culturally appropriate understanding of their meaning and function. Traditional healers offer an opportunity for doing this, and it would be a mistake to restrict their role to traditional illnesses as they may have credibility for preventing (and treating) 'contemporary' ills. Efforts to promote health must work through the norms of local communities rather than 'facing off' against them. Clinicians and those involved in the management of international aid should not assume that the consumers of health services in developing countries are as 'cognitively intolerant' as they may be themselves.

11

WHY NOT PAY ME?

The more one seeks to deprive the envious man of his ostensible
reason for envy by giving him presents and doing him good, the
more one demonstrates one's superiority and stresses how little
the gift will be missed.

(Schoeck, 1969, p. 21)

All too often . . . recipients trade their food stamps for alcohol,
Peace Corps members are pelted with stones, and foreign aid
recipients become alienated or hostile.

(Gergen, 1974, p. 187)

What is the best way to convey technology without someone
losing face?
(UN aid official, quoted in Gergen and Gergen, 1971, p. 100)

One possible first reaction to the Tanzanian case study would be to accuse the
host resignees, perhaps quietly to oneself, of being ungrateful. This accusation
would be based on the reasonable (and common) assumption that any kind of
well-intentioned help deserves appreciation and gratitude (Lerner, 1980). Perhaps this same basic assumption is implicit in the widespread attitude that many
host communities have developed a 'handout mentality'. In Fiji, for instance,
Traynor and Watts allege that 'Throughout the South Pacific, a "hand-out" mentality has been inadvertently created by aid agencies' (1992, p. 73). In an Indian
context, Sinha describes how recipients of aid are predisposed to 'an attitude of
dependency on some external agency' (1990, p. 89). In both cases, the implication is that people can only be roused to help themselves through the provision
of some outside benefit or incentive (Shaw and Clay, 1993). In short, they have
become indolent.

This could be donor bias (see Chapter 2) and Theory X (see Chapter 6) all over again, but writers still make such attributions and still become galled when doing so, possibly because they grate with what they, like many others, implicitly assume is the *raison d'être* of international aid. That core purpose is often seen as helping people to help themselves. Indeed, the notion of aiding communities to develop themselves has arguably become the ethos of many development efforts, geared as they are toward 'empowering' those most in need of help. Even here, however, one can detect the whiff of a possible fundamental attribution error: The insinuation, albeit unintentional, that the needy *could* have helped themselves, if only they had not been so 'underdeveloped' in the first place. Thus, there remains a pervasive tendency for donors in the West to 'expect that the recipient of their philanthropy will be duly appreciative' (Fisher *et al.*, 1981, p. 368). Aid donors crave reciprocity without realizing that recipients of aid may crave it too!

In this chapter, we argue that the integrity of the 'hosts at home' system (see Figure 1.1, p. 13) can become threatened by an unrelenting (and sometimes unwanted) 'assault' of unreciprocated gifts. Such gifts, and the attempts at reconciliation and social equity that they partly represent, may in fact merely highlight the injustices they seek to repair. Under such humiliating circumstances, the assertion that one will only *exchange* one's 'participation', i.e., for some kind of recompense, is perfectly intelligible, especially if there is a perception, on the part of hosts, that much aid is self-serving rather than serving the recipients of aid (Fisher *et al.*, 1981). As we shall see, such reactions are predictable, and explicable, on the basis of much social-psychological research and theory.

Logically enough, that same literature suggests that more reciprocity ought somehow to be introduced into the aid equation. This would mean a certain tempering of the idea, endorsed in many aid quarters and to some extent in this book, that hosts should dictate the terms of aid from the vantage point of their own needs. The most obvious candidate for allowing more reciprocity would be to place international aid on a negotiation and bargaining footing (for example, Carnevale and Pruitt, 1992; Kramer and Messick 1995; Pruitt, 1981; Rubin and Brown, 1975), but this would imply some kind of return to tied aid. Our analysis thereby identifies something of a conundrum with respect to rendering aid projects more respectful of the host community.

The 'Pay Me!' in Malaŵi

The following summary of data recently gathered in Malaŵi challenges directly the prevailing ethos of aid as helping people to help themselves. The findings indicate that so-called aid 'recipients' may demand financial remuneration for being aided. As we shall see, however, this 'affront' to the prevailing ethos of aid is not at all what it first appears to be.

During the 1992 drought and famine that swept sub-Saharan Africa, various anecdotes circulated around the town of Zomba. These anecdotes described how

some people allegedly responded to incidents where aid organizations had provided resources for support and relief. We realized at the time that these stories
could have some significance, maybe as sources of information about the way
that intended receivers of aid perceived their relationship with the would-be
givers of aid.

The anecdotes themselves were varied. They included a story about how the
local town community in Zomba had refused to unload its own consignment of
emergency food aid unless wages were paid for unloading the trucks. There was
also the story of a nearby village community, which made the assembly of
donated well-equipment conditional upon receiving paid wages to do so.

These two stories focused on drought-precipitated aid relief, but there were
other anecdotes relating to aid generally. According to one account, academics
often stayed away from aid-funded conferences unless the per diem (daily allow-
ance) was big enough to make a profit. Another anecdote, even more surprising
at the time, recounted how schoolchildren had declined to complete for free,
questionnaires designed to help improve the quality of their own education.

Such anecdotes, if true, would suggest that donors and recipients were at
cross-purposes in their respective definitions of aid and development. Some
Malaŵian communities, whether town, rural, or educational, appeared to be saying to the aid agencies concerned, 'Pay us to help ourselves!' In order to test this
'Pay Me!' hypothesis, we therefore presented the four anecdotes, in the form of
scenarios, to a sample of undergraduates at the University of Malaŵi (Carr,
MacLachlan, Zimba and Bowa, 1995b).

Many of these students had direct experience of the types of aid that had been
described in the stories. Moreover, many of them would themselves have been,
either directly or indirectly via the extended family, 'recipients' of such aid. We
therefore asked these recipients of aid projects to predict, for each type of
scenario, how the local community would probably react. The options we
gave ranged from 'Pay Me!' (for example, the town residents would refuse to
unload the maize unless they were paid to do so) through to self-help (for
example, the residents would gladly unload the supplies), with a midpoint for
'uncertain'.

Averaged over all four scenarios, 20 percent of these Malaŵians' choices predicted that the local community would react with 'Pay Me!', with a further 10
percent being 'uncertain'. Analyzing these various response rates by type of scenario, the academic vignette produced the highest estimates of 'Pay Me!' (32%,
plus 17% unsure), followed by the story of food aid to the town (27%, 5%),
then the school scenario (14%, 6%), and lastly the rural community setting (11%,
7%). Thus, the communities that had probably had the greatest amount of contact with Western aid agencies were also seen to react most clearly with 'Pay Me!'
– a trend that turned out to be statistically significant.

We also asked the respondents to explain their predictions. Why would those
local communities predicted to react with 'Pay Me!' responses do so? The reasons
given seemed to us to be perfectly reasonable and rational. Some payment would

more fully alleviate dire needs; galvanize what Johnson (1982) terms 'free-riders' who would otherwise shirk responsibility; reciprocate Western economic values (for example, 'nothing for nothing'); and/or give some clear and tangible benefits to the recipients. The last two reasons in particular suggested a certain tit-for-tat, and we started to realize that the 'Pay Me!' could be understood as an attempt to negotiate more social equity into the aid relationship. It might partly represent an attempt to close the power differentials between a powerful and yet remote donor on one side, and an otherwise 'helpless recipient' on the other. That, after all, in the final analysis, is what aid is *supposed* to be about.

The 'Pay Me!' elsewhere

Before exploring this motivational possibility further, the cynic might ask, can the Malaŵian data be explained more parsimoniously (and exclusively) by the idea of *Homo oeconomicus* (simply maximizing one's own profit)?

In 1997, the government of India asked Coca Cola to hand over majority control of its manufacturing process (Leung and Wu, 1990). Coca Cola offered a compromise, agreeing to cut its equity to a minority share but insisting on the retention of manufacturing rights in order to protect its unique formula. The Indian government refused this compromise, despite a costly withdrawal of the soft-drinks giant from the country. According to Leung and Wu, this tough stance was 'motivated by an ideological desire to downplay the role of foreign firms in the Indian economy ... If the Indian government valued economic pragmatism more than nationalistic ideology, the conflict would have ended once Coca Cola was willing to give up its majority ownership' (p. 213).

According to Cottam, host governments in developing economies may be sensitive to any multinational activity that symbolizes neocolonial hegemony, particularly with domestic audiences looking on (1989, pp. 452–3). After seizing assets belonging to the US multinational International Petroleum Company (or IPC), the Peruvian government refused to pay compensation, even under a threat that all US aid to the entire nation would be withdrawn. Thus, even when both trade and aid were at stake, maintaining a positive sense of cultural identity prevailed, a point that is echoed in Chapter 3 and our discussion of countries' refusing IMF loans.

Similarly perhaps, independent Indonesia severed aid links with its former colonial masters, the Dutch. In a rural community setting in East Africa, Porter *et al.* (1991) describe how preserving socio-cultural values and identity (for example, land security, preserving traditional institutions) was often more important than meeting so-called 'survival' needs (for example, piped water). Cox (1995) has argued that nationalism and tribalism are rising not falling, while Tajfel (1978) found that groups often prefer to take a cut in material reward, in order to position themselves favorably in comparison with an outgroup. In Malaŵi, we found that even during the drought of 1992 people did not *passively* accept aid, but instead asserted the right to payment for their labor. Anecdotally, too, we have seen that aid projects often produce 'Pay Me!' reactions, although the cases themselves are rarely documented empirically.

One exceptional case concerns a food-for-work project in rural Ghana (Whaples and Ogunfiditimi, 1979). This was a 'self-help' project using food as a 'work incentive' (p. 291). The authors of the study interviewed 400 such 'workers', factor analyzing their responses to a number of attitude questions. In addition to concerns about security and tradition, one of the factors that emerged from this analysis was 'incentive', defined as 'attitudes toward types of remuneration for services performed or to be performed' (p. 292). Statistically speaking, the emergence of this factor means that there was a wide range of opinions on the issue. As in Malawi, a significant proportion of people must have endorsed the 'Pay Me!'

More recently, Kaul (1989) has conducted a study of eleven development projects, spanning Asia (Bangladesh, India, Sri Lanka) and Africa (Botswana, Tanzania, and Zimbabwe), on rural credit to urban housing. In addition to features like genuine recipient participation and goal definition, both recipients and staff required some material benefits; some 'income generating mechanisms' (p. 19). Without these tangible benefits becoming ends in themselves (people wanted income-generating *mechanisms*), project success was contingent on respecting a wish to be 'paid back' for something.

In order to determine what that 'something' might be, we need to try and step inside the shoes of an aid project recipient.

How does it *feel* to be an aid 'recipient'?

This basic question is actually rarely considered, at least compared with its corollary of why people *give* (see, for instance, the differing proportions of literature in the review by Bierhoff and Klein, 1990). That bias partly reflects perhaps the comparative wealth, and ability to give rather than to receive, of those countries wherein most of the literature concerning aid has been both produced and published.

The social-anthropology literature, however, has suggested that receiving aid may be aversive. For example, when money is given as a gift, it may be more acceptable when appropriate to the level of intimacy and relative status between donor and recipient. Otherwise, gifts of money and nothing else can transmit offensive (for example, patronizing and demeaning) messages (Burgoyne and Routh, 1991). Similarly perhaps, Krishnan and Carment (1981) found that Canadians were less accepting of help when donors offered *larger* amounts of aid. In many non-Western societies, non–monetary gifts are used to signal respect for traditional social structures (for example, Silverman and Maxwell, 1978). From a host's perspective, therefore, money and gifts, and even aid generally, could become symbolically threatening.

In psychology, there exists a potentially useful experimental literature on individual differences in reactions to help and assistance (for reviews, see for example,

Bierhoff and Klein, 1990; DePaulo *et al.*, 1981; Nadler and Mayseless, 1983). This literature has focused on the importance of self-esteem, which we have already argued may be related to pride in the *collectivity*. The literature itself (to summarize briefly) makes the point that pride will mediate the effects of aid. On important issues, high self-esteem has been linked with greater resentment of the donor, a stronger desire to somehow pay back the help, and greater readiness to self-help. For low self-esteem cases, when help is felt to be overwhelming, people may get locked into a vicious circle of increasing passivity and dependency. Thus, the overall message for aid might be that it is important to build into projects mechanisms for maintaining collective esteem, as well as reciprocity.

Along with others in development studies, we have argued that a great deal of social science, both in the West and in developing countries, points to the conclusion that social and organizational change that is felt to be imposed is unlikely to sustain (Ingram, 1994). Such impositions are likely to constrain a community's sense of cultural identity, thereby motivating the group to reassert that identity through resistance of various kinds. International observers who perceived aid as more manipulative and self-serving appreciated its value, and integrity, less (Gergen and Gergen, 1974; Nadler *et al.*, 1974), while developers in Kenya ultimately paid the price for assuming that cultural traditions could be overridden in the name of meeting more so–called 'basic', *self*-preservation needs (Porter *et al.*, 1991, p. 149).

In sum, a variety of available evidence suggests that there may be an alternative to the implicit benevolence of 'helping people to help themselves'. This alternative might somehow entail a more balanced game or narrative script for aid (see Chapters 8 and 9), one in which there is a mutually respectful *exchange* of (a) materials and money for (b) participation and labor. That might lift some of the social pressure from the local community, especially perhaps if its members traditionally associate well-being with autonomous productivity, as rural communities often do. What we need here is a theoretical mechanism for understanding *why*.

A norm of reciprocity

The social-science literature contains at least one behavioral principle to explain and support the idea of making aid more of an exchange. A key construct in much social science is that of a 'norm'. Norms are essentially unwritten group standards specifying (in)appropriate behavior, beliefs, and feelings. A moment's thought here may convince you that all societies have them (social norms), and that they function to make the world a predictable and perhaps more tolerable place in which to live.

In a classic study by Sherif (1936), described in detail here in Chapter 3, people were placed in a darkened room in which a small light appeared to move erratically. They quickly established a group norm about the distance it was moving on each trial, and later applied that group standard even when judging the stimuli by themself, for up to one year afterwards (Sherif, 1937). Since the original study, as we saw in Chapter 3, such tendencies for groups to establish norms has been widely supported (Wilke and Van Knippenberg, 1990).

These studies tell us something fundamental about social influence. People tend to gravitate toward the average of their individual judgments. As today's postmodernists might reiterate, Sherif's central point was that *reality is sometimes socially negotiated* (see also Festinger, 1950). The same type of reaction has also been recorded in classic studies of social pressure (for example, Asch, 1956) and social innovation (for example, Moscovici, 1976, 1980). In each case scenario, people finding themselves at loggerheads spontaneously looked for a position that symbolized respect for both parties. Social interactions were often characterized by what has been termed a 'norm of reciprocity' (Gouldner, 1960).

Like other social scientists in more recent times (for example, Triandis, 1978), Gouldner believed that this norm is universal (see, however, Mathur, 1983). We prefer, however, to think of it as a 'meta'-norm, i.e., one that influences many social situations, in a great variety of cultural contexts. Some diverse examples would include: cultural gift-exchange rituals; the Confucian ethic of repaying kindness with kindness and evil with justice; Kant's 'categorical imperative' (act in the way you wish others to act); evolutionary arguments that reciprocal altruism is adaptive; cross-cultural psychology experiments in which transgressions of reciprocity are punished with social disapproval (Morris *et al.*, 1995); and the cross–cultural finding that relationships seem to be based on a principle of social exchange (Fijneman *et al.*, 1996).

In relation to aid, a field study described by Morris *et al.* (1995) describes how the Hare Krishna technique of pinning a gift on to passers-by, and then asking for a donation, left donors annoyed but nevertheless unable to refuse the request for aid. They felt obliged to reciprocate. In terms of expatriate–host relationships, mastering a host's language may represent an unwelcome intrusion (Mamman, 1995), whereas Pidgin can function as a tolerable halfway house (Carr, 1996c). In Chapter 7, we saw how important it was for aid-funded scholars to feel that they were getting a fair deal. Similarly perhaps, among rural communities in Malaŵi, reciprocity has contributed much toward the success of fish-farming aid projects (Mills, 1994).

Even the so-called 'universal' *exceptions* to the reciprocity norm suggest its relevance. According to Leeds (1963), these are people who are in some respects less able to reciprocate, such as infants and the infirm. Remembering the popular dictum for aid, and its possible insinuation that recipients have, until the aid arrives, been unable to help themselves, they (the 'recipients') would be cast in the same kind of role as infants and the infirm.

What the reciprocity norm explicitly says about the human factor is that people are often *motivated* to retain reciprocity. The corollary of this, of course, is that they will be motivated to *restore* reciprocity whenever it has been denied. In the relevant literature, there are at least three basic, and often complementary, reasons for that restoration process (Fisher *et al.*, 1981). These are: threats to self-esteem (pride); reactance (Brehm, 1966; see also Chapter 1); and the desire for equity restoration (see Chapter 6).

A prediction made by all three theories is that unreciprocated aid will motivate recipients to derogate the donors or the aid itself, whether to maintain self-esteem, to reassert autonomy, or to balance outcome-to-input ratios. In Chapter 3, we learned how reciprocation is frequently blocked by the rigidity and linearity of the aid chain, meaning that aid cannot be reciprocated. It is therefore unfortunate that extant research, both in the West and within some developing countries, has tended to support the theories' prediction that lack of reciprocation *will* cause the aid process to sour (for detailed discussions, see Fisher *et al.*, 1981; Gergen and Gergen, 1971).

In one Western study, for example, undergraduates were helped by a confederate of the experimenter (Shumaker and Jackson, 1979). In just one session, those recipients who were denied the opportunity to reciprocate readily derogated the confederate, even when they were given the opportunity to help a third person. Nothing less than literal reciprocation would do. In another study, help was given with the remark, 'this must be hard for you' (De Paulo *et al.*, 1981). Recipients were divided into those with high and low self-esteem. Compared to those individuals with high self-esteem, the negative signal had less of a motivating effect with regard to a difficult task on those with low self-esteem. Insofar as collectivism emphasizes modesty before the group rather than 'self'-esteem (Smith and Bond, 1993), unreciprocated and essentially patronizing benevolence may thus become relatively demotivating. From this viewpoint, receiving aid reduces one's sense of dignity, implying the role of beggar (Gergen and Gergen, 1971, p. 95).

Several studies have focused specifically on international aid. In one of these, sixty donor and host officials, working in various international assistance projects and representing twenty-two countries, were interviewed about the importance of various factors in the aid process (Gergen and Gergen, 1971). Perceived respect for the recipient country proved to be the most important factor in moderating these aid workers' reactions toward, and acceptance of, aid. This included, too, failing to respect the *autonomy* of the host, a disrespect that may provoke a certain reactance, or contrariness, on the part of the host.

Similarly, Gergen and Gergen (1974) analyzed survey archives of literally thousands of citizens, from thirty-seven different developing nations, concerning aid received from Western donor countries. As in the 1971 study, reactions to donors who were perceived as imperialistic (including toward their own peoples) were comparatively unfavorable. Aid officials reported that allowing the recipient to reciprocate was essential for aid project success (Gergen and Gergen, 1971).

This resonates somewhat with our own experiences in Malaŵi. There, aid workers were frequently perceived by hosts not to be altruistic or giving (this is actually often the case among Westerners, see Fisher *et al.*, 1981). Instead, they were often seen to be 'milking the system', whether by drawing large international salaries, possessing four-wheel drives, or sojourning in luxurious hotels. Being surrounded by the good fortunes of others has long been linked with feelings of relative deprivation (Stouffer *et al.*, 1949). Furthermore, as Fisher *et al.* (1981) caution, aid that is seen to be easily affordable may leave recipients feeling less obligated toward the donor, as well as less inclined toward self-help. It is therefore not surprising that one of the major explanations for the 'Pay Me!' in our Malaŵian setting was 'nothing for nothing'.

The other side of the Malaŵian 'Pay Me' was seeing some purpose in spending one's labor and time working on the project. The benefits of aid were not always clear to the intended 'beneficiaries' themselves. In the words of some of our Malaŵian student respondents, 'People have other needs apart from the ones that aid projects can satisfy!' Similarly, following a presentation on the 'Pay Me' to an international development studies audience (Carr and Munro, 1994), a Papua New Guinean 'recipient' of forestry aid, asked to explain his reluctance to 'participate' for free, reportedly retorted: 'Working on a forestry project would keep me from my everyday subsistence activities, so Why *not* pay me for [i.e., reciprocate] my time and trouble?'

Breaking the circle

The norm of reciprocity also has it that money itself is not the real issue. So donors who deem aid recipients as helpless and incapable of reciprocating may be *establishing* a norm by paying them to participate. From a dynamical systems viewpoint, this may then set up expectations among recipients, and a perception that aid projects represent a source of gain or income for those involved. If this expectation is met the next time around, a 'Pay Me!' culture is reinforced. At the next aid project, both the perception of helplessness and the expectation of payment are higher, and set to spiral. Figure 11.1 pictures this process of perceived need gradually deteriorating into outright 'Pay Me!' Aid officials the world over have in the past reported that a sustained flow of aid eventually results in a culture of 'entitledness' (Gergen and Gergen, 1971). Systems analysis suggests one slippery slope, paved with good intentions, by which that particular destination may be reached.

The process and outcome painted in Figure 11.1, as well as the suggestion that money itself is not the real issue, are borne out in project evaluations where recipient criticisms frequently focus on issues such as per diems being too low, no travel allowances, accommodation not suitable, etc. When one of us was conducting community research on AIDS in Malaŵi (McAuliffe, 1994b), and inviting people to participate in focus-group discussions lasting two hours, colleague researchers said that every participant would have to be paid 50 Malaŵian Kwacha (about US$15) for their time. This was refused, on the grounds that it might bias responses, but instead the group facilitator offered to answer any questions the participants might have had about AIDS (i.e., reciprocate) at the end of the discussion. The money never became an issue. The circle was broken.

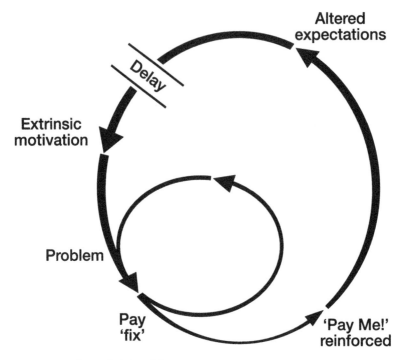

Figure 11.1 A pay 'fix' that fails
Source: Adapted from Senge (1992).

How does reciprocity link with the 'Pay Me'?

Turning back now to the first signs of 'Pay Me!', and their explanation, we have seen how aid 'gifts' are often responded to with requests for payment to facilitate the use of these gifts (for example, unloading and assembling well-equipment during a drought). Superficially, it may seem that a request for payment is following the receipt of a gift, and that this does not therefore constitute reciprocity. However, if the provision of aid is seen as undermining its recipients (who *could not solve the problem on their own*), then aid no longer has the status of a gift. Instead, it may be experienced as *taking something away* (valuable time, self-respect, collective autonomy, pride in traditional culture, etc.) from the recipient.

In such a context, the norm of reciprocity demands that those who have taken away something should somehow *pay* for it. In this way, aid gifts must be counterbalanced by demands which will restore some sense of equity into the transaction. If the reciprocity norm is as robust as we think it might be, *many forms of aid will lead, eventually, to 'Pay Me!'* By the time that emerges however, it may have already soured the donor–host relationship, a deterioration that could be exacerbated if the donor then demotivates the recipient still further by introducing unnecessary payment (a process already discussed at length in Chapter 6).

Yet there is another possibility, suggested by the results of a cross-cultural experiment (Gergen *et al.*, 1975). In this study, donors who sought *an equal return* were liked significantly more than those who demanded either too much in return for helping a needy colleague (exploitation), or *nothing at all* (benevolence). Similarly, Nadler and Mayseless (1983) report that providing an opportunity to reciprocate allows a way to preserve and restore pride and intergroup relations. These in some respects counterintuitive findings are depicted in schematic form in Figure 11.2. Striking such a balance could become especially salient in *bi*lateral aid projects, where there is a readily discernible benevolent donor, and therefore a clearer threat to the recipient group's pride and autonomy (Fisher *et al.*, 1981).

What, then, might count as a return that is 'equal'? In his aid-project review, Kaul (1989) has argued that reactions similar to 'Pay Me!' can be managed very effectively, by offering *mechanisms* for reward. These, by implication, have to be earned and thereby 'give something back' to the project, rather than becoming financial rewards in themselves. The examples reviewed by Kaul include repaying credit loans; building a house; learning by doing; and earning some kind of

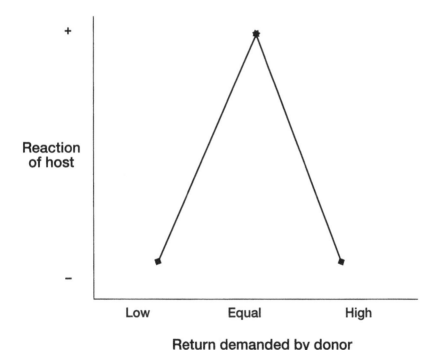

Figure 11.2 Optimizing reciprocity
Source: Adapted from Gergen *et al.* (1975)

195

genuine career prospects (for project staff). In the psychological literature, one particularly interesting factor that reduces any dislike for a helper and increases the readiness proactively to ask for aid when needed in the future, is opportunity to *pass on* benefit, by helping a third party (Castro, 1974). This finding raises the intriguing possibility that cultivating an ethos of helping people to help *others* is one way of managing 'Pay Me!', a possibility arguably supported by the concept of 'trickle down' (see Chapter 8).

Overall, the evidence 'may be interpreted as a plea for more reciprocity in altruistic relationships . . . Altruistic responses which are embedded in a mutual give and take may have more desirable effects than one-sided [exploitative or benevolent] aid which offers no opportunity to repay' (Bierhoff and Klein, 1990, p. 260). This suggests a principle within which an appropriate sense of reciprocity can be allowed to develop: through negotiation and bargaining.

Kaul's (1989) review found that a negotiatory framework could be applied at multiple levels in the aid process, or 'aid chain' as we and others have termed it (see Chapter 3). For example, there could be negotiatory interfaces between sections of the aided community itself; between lower and higher officials; between NGOs and governments; and between beneficiaries and project organizers (for example, in the setting of goals and evaluation of outcomes). However, we believe that aid agencies need explicitly to signal *permission* for a local community to engage in a negotiation process. While everyday skills of marketplace negotiation may be very familiar to the hosts (and often quite foreign to donors, suggesting where *they* might require training), the (formal) negotiation of *aid intervention* contracts will often be quite unfamiliar to hosts.

In this sense, hosts may benefit from training programs which explore the complex dynamics of the donor–recipient relationship, and in doing so increase their awareness of the roles they may adopt in negotiating aid intervention. Historically, we might say that the recipients of aid have been 'positioned' (see Chapter 8) by their donor partners into playing the wrong 'game' – in which recipients are expected to be grateful beneficiaries of donor gifts. According to this argument, partners in aid need to reposition themselves such that recipient communities can *give* their acceptance to negotiated terms, and thereby reciprocate.

Standing back from – meta-positioning – relationships in which one is deeply self-interested is terribly difficult, but we believe that this can be achieved through appropriate training techniques, as mentioned above. We now consider just one example of how this can be done, in the form of ultimatum games (Guth *et al.*, 1982). Rather than focusing on economic competitiveness (Beardsley, 1993; Carnevale, 1995; Nash, 1950), this paradigm has concentrated much more on social relationships between 'donors' and 'recipients' (Kramer *et al.*, 1995).

In an ultimatum game, one partner is empowered to propose a division of money while the other side must accept or reject the division. If the division is rejected, then both sides lose out. For our purposes, such scenarios clearly have certain parallels with international aid. Moreover, in Western studies (for example, Murnighan and Pillutla, 1995), most 'donors' tend to treat the

196

situation as an opportunity for material gain, which is precisely what Western aid donors are increasingly being accused of in some quarters.

As you might expect, too, the 'recipients' in these simulations tend to define the situation in moral terms, eventually rejecting the donors' self-serving 'gift'. Even from a Western recipient's point of view, moral considerations tend to over-ride purely economic ones. In systems terms, and as suggested in Chapter 4, such asymmetry then fosters mutual unhappiness. Donors increasingly feel that recipients are ungrateful, while recipients increasingly feel that social injustice has been done. There is an escalation of social disaffection and conflict.

The resonances here with aid are quite striking. We have seen how donors sometimes perceive hosts' 'Pay Me!' reaction to reflect ingratitude, even though it may actually represent an attempt to establish some kind of moral and social equity with the donor; as well as how attempts at a quick pay 'fix' can increasingly begin to backfire on the donor. Ultimatums consistently fail, or produce negative 'Pay Me!', because they ignore the social dynamics of reciprocity. They are thereby felt to be socially unjust. Often, too, and adding injury to insult, they are plainly incorrect. As one African leader said of aid-funded coercive democracy, 'the Western multiparty systems were a *result* of prosperity not the cause' (Gertzel, 1996, p. 3, emphasis added). In fact, archival social assessment indicates that individualism tends to follow economic growth, not the reverse (Hofstede, 1984; Hofstede and Bond, 1988).

Assessing cultural contingencies

Ultimatum games have been examined in a variety of cultural contexts (Roth *et al.*, 1991). In the latter study, there were large differences in the amounts that recipients found acceptable, with Japanese and Israeli players being willing to accept lower amounts than Yugoslavs and Americans. Individualism (vs. collectivism) is often correlated with being comparatively insensitive to the balance in one's relationship with a negotiator (Chan *et al.*, 1994; Leung and Wu, 1990; Smith and Bond, 1993). Thus, the research indirectly suggests that there could be wide variations in the 'Pay Me!' (and reactions to it) according to the cultural mix of parties involved in a particular aid project.

Ultimatum games are also one example of how recipient communities can be provided with training which might increase their awareness of difficulties in the aid process. We are not suggesting that donors should – or even would be able to – 'train up' aid recipients in terms of negotiation skills by whatever type of nego-tiation game! When it comes to skills such as negotiation, it is self-evident that many aid recipients have these in abundance, and that they use them every day in the marketplace of life in developing countries (Rugimbana *et al.*, 1996). But by appreciating some of the dilemmas of aid from a meta-perspective, local com-munities might more easily pass beyond 'seeking permission' to negotiate aid contracts; toward giving them*selves* permission to use the full range of their own skills. In short, *to play their own game.*

Concluding remarks

Overall, the theory, research and recorded practice, consistently point to one overarching conclusion. It is imperative not to be perceived as coercive, whether by outright dominance or through a more subtle form of dominance, namely the benevolence of 'helping people to help themselves'. Such heavy-handedness is bound to lead, eventually, to a (re)surfacing of socio-cultural resistance and conflict. Like much social psychology, the empirical evidence from the aid arena supports the principle that the more one constrains (whether wittingly or no), the less one can sustain. A *mutual* sense of reciprocity is evidently required, and negotiation provides a framework within which that sense can potentially develop.

12

MINIMAL CONSTRAINT

There are wholes the behaviour of which is not determined by
that of their individual elements.

(Max Wertheimer, founder of Gestalt psychology)

We learn best from experience, but we never directly experience the consequences of many of our most important
decisions ... Cycles are particularly hard to see, and thus
learn from.

(Senge, 1992, p. 23)

Synopsis

The message that emerges from this book is that there is one important synthesizing 'human factor' in international aid, namely the motive toward social
equity, as defined in context. Despite globalization in general, and aid for trade in
particular, there remains a primary, perhaps essentially a-economic, role for this
particular human factor (Porter *et al.*, 1991, p. 149). Ironically, the nemesis of
some aid projects may be the very social injustice and inequity that they seek to
correct. Based on local definitions of social justice, *social inequity leads to project
entropy.*

Minimal constraint

The implications of social equity become clearer, perhaps, once we consider what
it might mean in terms of respecting social and/or socio-cultural identity. As we
have argued at several points in this book, the question of social (as compared
with individual) identity is likely to be relatively (but not exclusively) salient for
people in comparatively collectivistic societies. We now suggest that minimal
constraint is a key to retaining social identity in international aid. *Aid should not
be constraining.* This is because social identity is something that is implicitly but
undeniably brought to, and laid upon, the aid table. It is always there, and it is
always placed in jeopardy in any form of project to 'aid' other people.

199

A retrospective glance at 'The case' in Chapter 1 reveals that, from the beginning, the Tanzanian hosts (or at least those 'at the coalface' rather than in central government offices) were never consulted about the need for, and form of, the aid they were about to 'receive'. Viewed in terms of our principle of non-constraint, a mistake was made as soon as the aid agency unilaterally decided to 'send an expatriate' to its chosen field setting. From the outset, the recipient community was constrained and humiliated into a disempowered and passive role. In Craig and Porter's (1994) terms, their subjectivity was in danger of being engulfed, unless some form of 'disengagement' could be found.

In Chapter 2, we introduced social psychology, in particular attribution theory. The central plank in this theory is motivational, namely a need to make sense from one's experiences. Throughout that quest, people seek an ego-preserving balance between their cognitions, especially those that contain knowledge about themselves. Anything that disturbs this balance tends to precipitate some kind of ego-defensive maneuver, from switching off metaphorically (or literally), to blaming the victim. Aid campaigns that rely on projecting human suffering 'in your face' may in fact be *constraining* sizable segments of the public from donating. What they may be doing is causing an affront to the quasi collective identity of their viewer audiences; precipitating cross-cultural stereotyping that is unconsciously 'designed' to restore the balance between 'their' outcomes and inputs, and 'ours'.

In Chapter 3, we extended some of these psychological mechanisms to control orientation in aid organization decision-making. When aid projects start to go awry, for instance, blame may be conveniently shifted down, on to the next link in the chain, which, or who, is ultimately the host organization or community (Gow, 1991). The whole function of a chain is constraint, and the aid chain, from a social-psychological point of view, is probably no exception. The differing agendas of the various links in the chain constrain the development of an agreed and appropriate set of objectives for aid projects. Lack of coordination and poor communication are further constraints on aid effectiveness. Power differentials between aid workers and aid recipients prevent true participation by communities in aid projects designed to benefit them.

A clash of cultural values is another potential source of perceived constraint. In Chapter 4, we learned how so-called 'democracy', and even aid itself, can be construed as value laden. Culture is not a fragile entity, and it would be naïve in the extreme to believe that one set of such values (for example, those of a donor agency) will ever eclipse another (for example, those belonging to a host). Quite the contrary in fact. As the former Under Secretary to the United Nations and Director of the World Food Programme has warned, in relation to rising fundamentalism and other assertions of cultural identity, 'Imposition from abroad is to invite an eventual counter reaction' (Ingram, 1994, p. 61).

At the individual level, this tendency to react contrary to perceived constraint is termed 'reactance' (Brehm, 1972), while, at the level of (sub)cultural groups, Tajfel (1978) has argued, using many examples from developing countries,

200

that assimilationism will eventually backfire and give rise to anti-conformity and cultural revivalism. Somewhere in-between, and in the context of an expatriate aid assignment, we gave a systemic view of how differences in cultural values could lead to escalating misperceptions on both sides. The essence of the conflict, in both cases, is that not enough respect ('face') is seemingly being given. Underneath that, however, is the more fundamental problem that each side is defensively reacting against what actually feels like duress.

Chapter 5 introduced the notion of incremental improvement, where change is effected in small steps rather than large leaps. However, what is crucial in this process is that aid workers learn from the communities they seek to serve. Thus, once again, the recipients of aid cannot be constrained by pre-determined agendas. In focusing on the process of change rather than its outcomes, incremental improvement is about getting relationships right and seeking a scale of change which can be incorporated into the matrix of social structures and institutions. We have argued that such change is necessarily going to be slower but surer.

As we saw in Chapter 6, expatriates under foreign-aid contracts are not exempt from certain constraints either. The essence of the verb 'to constrain' is to use an excessive level of incentive, and this definition probably fits the way that certain aspects of aid salaries may appear to different expatriates. The more altruistic and socially sensitive expatriate will probably find the *comparative* aspects of his or her salary *morally* constraining, whereas the same comparative remuneration may begin to get a hold of the more extrinsically motivated expatriates, to the extent that they become less and less willing to 'work themselves out of a job'. In each case, the person affected, whether consciously or not, may begin to develop an inflated opinion of her or himself compared with local colleagues, thereby increasing the risk of project failure. These are the human-factor dynamics of relative privilege and gratification.

Meanwhile, the issue of who should adapt to whom, as raised in Chapter 7, is inherently about constraint, for example, through the themes of perceived prejudice and exploitation. Most importantly of all, however, we learned that those particular feelings may bear a close relationship to negative recommendations about the donor country later in life. Thus, socio-economic factors, such as background and income, were outweighed by social-psychological ones, in particular by attributions that universities in the host country were 'out to make a quick buck', and did not really care about the welfare of its clients. In that sense, the system was too rigid to accommodate to their needs. Particularly for the aid-funded students, who perhaps had less 'voice' than those who were full fee-paying, the system was *constraining*.

The prospect of external or internal exile, discussed in Chapter 8, is a continuation of the same theme. Here, we learned how the curriculum, and the system on return home, is often driven by concerns that are alien to the broader community. Having spent years 'climbing to the top' of a foreign system, one has a vested interest in not decrying that system, and in working within it. But that is

PSYCHOLOGY OF AID

so only to a certain extent, because one is thereby constrained into a form of exile from one's own community and its own development needs. Understandably, therefore, the result can be a certain ambivalence and apparent disengagement from the system – including perhaps the aid agencies that support it.

If an aid scholar or trainee does decide to return home, he or she may be required to meet the added constraints imposed by a foreign work ethos, such as being competitive and self-promoting (see Chapter 9). Added to this, they will be obliged to work not in a traditional manner, which might center on highly cohesive work groups, but instead to work alongside and below members of outgroups. The resulting intergroup dynamics may therefore include *inverse* resonance, meaning that an otherwise positive human factor (ingroup loyalty) has been 'corrupted', through inappropriate and alien ways of organizing work, into a 'negative asset'. Little wonder, perhaps, that some outside observers complain that managers in sub-Saharan Africa 'lack commitment' to job roles.

The chapter on tolerance and development indicated that constraining people to accept one particular form of health care, so-called 'modern medicine', may have no real psychological foundation or support. Consumers of health services tend to resist the imposition of any monoculture, especially if it emanates from a foreign culture. Like aid itself, 'modern medicine' too is a form of culture, and those aid projects that have the foresight to be pluralistic, thereby giving more face to their customers and their trusted healers, are more likely to be successful.

This need for showing face is brought into focus most clearly of all perhaps through the 'Pay Me!'. Our discussion in Chapter 11 focused on the possibility that the culture of aid is inherently demeaning to many of its recipients, primarily because it often fails to accommodate a powerful and frequently compelling social norm. Reciprocity is the natural enemy of constraint. It acts to restore pride, including group pride, by showing that the 'recipient' has something to give as well. True negotiation would seem to be an essential framework for introducing more reciprocity into the aid process, at the expense of constraint, if each side manages to 'recompute' what it actually brings and can bring to the table.

Closing the circle

We began this book with the suggestion that systems theory, and the social-interactive view, could be a way forward toward a better understanding of the human factor in international aid projects. That systems perspective gave us a framework for organizing the material, and for beginning to think about how the different human factor components in the system might fit together, in a functional (rather than prescriptive) sense. Having examined each of the functional subsystems in turn, as well as some of their linkages, we conclude by integrating them into an overall 'gestalt'.

Throughout the preceding chapters, two particular 'systems archetypes', or 'nature's templates' as Senge (1992) calls them, occur repeatedly.

One of these is the fix that fails, which we have seen operating in all four quadrants of Figure 1.1 (see p. 13), often through over-promising and under-delivering. Included (for instance) would be the 'promise' to donors at home that donations will make a difference; to TC expatriates that social equity will prevail; to aid scholars abroad that their education will be suited to work back home and therefore worth the investment demanded; and to hosts at home via payment-for-participation.

The remaining systems archetype that has recurred at several points in the book is escalation. We believe that this 'figure-of-eight' system, with its *two* vicious (or virtuous) circles, could be particularly important for understanding (and preventing) the development of social conflict, through international aid projects. In keeping with the theme of managing the dynamics of conflict across diverse cultural groups, the example we now discuss concerns the process of double demotivation, first introduced in Chapter 6, in regard to donors abroad, and then later in Chapter 9, about hosts at home.

We shall attempt to use this discussion to integrate the various systems that have been identified as operating in international aid work, thereby 'closing the loop' and defining the overall aid system at a social-psychological level. In themselves, systems archetypes are merely the functional elements, not the final process itself.

The whole point about escalation, as it applies to interactions between *people* and *groups*, is that each side believes it is merely acting defensively, whereas to the other side it appears to be behaving aggressively. Whatever the *actual* motives, this is a tragic case of perception becoming reality. In double demotivation, for instance, we have described how some expatriates, in defense of their ego, may begin to develop an inflated view of themselves as well as a deflated one of their counterparts. Host counterparts, in turn, may 'disengage' from work roles, as they attempt to withdraw some of their social input in protest.

From the point of view of the expatriate, however, this can look like plain surliness or unreliability. From Figure 12.1, any notions that the expatriate may have begun to have about being 'better' are reinforced! This air of 'superiority' can then be detected by the host counterpart, who may already be somewhat indignant at the salary gap. That results in more disengagement, and the vicious cycles soon begin to become amplified. As the expatriate progressively hogs more of the limelight, social conflict increases. If left unchecked, it may escalate into a destructive conflict between donors and hosts, such as occurred at the University of Papua New Guinea (Chapter 9).

Elsewhere, we have suggested that expatriates' disparagement of host counterparts may produce a 'Pygmalion effect' (MacLachlan, 1993b). Pygmalion, in Greek mythology, was a sculptor who believed so much in his own work that it came to life. TC is about education and training, and the Pygmalion effect is today widely recognized in education generally (Tauber, 1997). There, including in expatriate aid, it has come to denote the idea (and often the reality) that a teacher's expectations can sometimes become a 'self-fulfilling prophecy' (Merton, 1968; see also Chapter 6).

The systems-theory perspective, however, offers us an additional possible insight into how this process might function. From Figure 12.1, we now suggest that technical 'cooperation' could in fact be creating a *double* Pygmalion effect! Socio-economic contingencies could be conspiring with several 'human factors', giving each group's defensive maneuvers, as they attempt to justify the socially inequitable pay differential, an air of non-cooperation. The social tragedy of this scenario – its dynamic – is that these cross-positioning maneuvers are set to develop into social conflict or project entropy, for example, if one or more parties (as in 'The case' in Chapter 1) eventually exits the system.

In our view, the dynamic portrayed in Figure 12.1 is not an isolated example. As depicted here in Chapters 2 and 4, the archetype may apply to many different types of cross-cultural (and cross-positioned) social conflicts. *Within* cultural settings, we can envisage the same process unfurling between the 'haves' and the 'have nots'; in *multi*-cultural settings, between one ethnic group and the next.

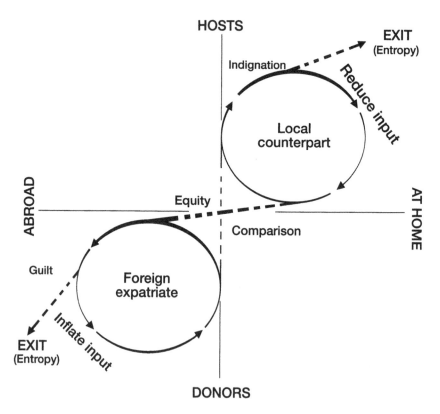

Figure 12.1 A double Pygmalion effect
Source: Adapted from Senge (1992)

Within *organizational* cultures, we can envisage the dynamic occurring, as in propagating Theory X vs. Y, or motivational gravity, between management and shop floor. These groups, after all, normally draw different salaries and have different opportunities, despite the 'lower levels' often regarding themselves as the equal of their 'superiors', while managers often see themselves as (superior) hero innovators (Senge, 1992).

In short, wherever one finds social inequity, one may also find escalation. If so, the system, *through the double Pygmalion dynamic*, may represent something of a 'template' for social conflict. Subsumed here, for instance, would be the corrupting and contaminating influences of power and powerlessness.

In the case of international aid projects, these interlocking and mutually sustaining cycles are portrayed in Figure 12.2. This superimposes the various systems discussed above, and in this book as a whole, on to and into one another, along with one or two other intuitively possible linkages, such as enmity developing between expatriate and host *abroad*. According to Senge, 'reality is made up of circles, but we see straight lines' (1992, p. 73). What Figure 12.2 is suggesting

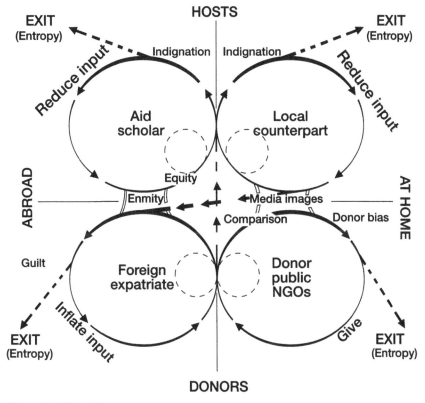

Figure 12.2 The aid system
Source: authors

is that international aid projects, at the level of human factors, may actually (sometimes) resemble a circle of vicious circles!

If there is one underlying message in Figure 12.2, and therefore in the book as a whole, it is this: just as neglecting the human factor may have created, and added momentum to, an 'aid cycle', so too a better understanding of the human factor, and systems dynamics, in particular, may eventually help to break or even *reverse* that cycle.

GLOSSARY

Aid chain Any linear and top-down communication channel between the various organizations in the delivery and receipt of aid.

Cognitive tolerance The ability to tolerate two or more different, and conceivably contradictory, views.

Convergence When a task is new and difficult, group members may tend to converge and coalesce on a position that reflects the average of the range of initial positions within the group.

Donor bias A theoretical tendency for Western donor publics to over-explain poverty in developing countries in terms of character traits (e.g., laziness) within the poor themselves, compared with what the poor themselves perceive to be important, especially situational factors (e.g., natural disasters).

Fundamental attribution error A tendency, found in many Western settings, for observers to over-explain others' behavior in terms of character traits, often overlooking equally or more important situational factors.

Incremental improvement The subjective impression of how much change would constitute a just noticeable difference or development in quality of life.

Inverse resonance When people are thinking and feeling as a tight-knit group rather than as individuals, this may give rise to outgroup discrimination. This is especially likely when the two groups are similar rather than dissimilar and therefore non-comparable.

Large leap From the context of a comparatively wealthy background, donors' expectations of what would count as a just noticeable, incremental development for the poor.

Motivational gravity Negative social reactions to personal success at work, both from threatened superiors ('Push Down'), and from resentful, or protective, co-worker peers ('Pull Down').

Norm of reciprocity A theoretical tendency for donors to expect gratitude from aid recipients, and for the recipients themselves to expect something in return for their investment in, or debasement through, an aid project.

Privilege proximity A combination of being relatively privileged as well as close to those less privileged than oneself, resulting in sharp feelings of guilt

207

and possibly blaming the victim as a means of reducing that guilt. Expatriates on comparatively high international aid salaries, for example, may begin to derogate the abilities of their lower-paid host counterparts, thereby partly justifying their higher pay.

Social equity   A motivational goal, implicitly or explicitly sought by both parties as groups in the aid transaction: to attain and retain a sense that each group's outcome is proportional to its input costs.

Transactional positioning   When dealing with one another, donors and hosts may implicitly adopt various roles within a socially constructed narrative or narratives.

REFERENCES

Aamodt, M. G. (1991). *Applied Industrial/Organizational Psychology*. Belmont, CA: Wadsworth.

Aamodt, M. G. (1996). *Applied Industrial/Organizational Psychology*. Belmont, CA: Brooks/Cole.

Aamodt, M. G., and Surrette, M. A. (1996). *I/O Psychology in Action*. Pacific Grove, CA: Brooks/Cole.

Aamodt, M. G., and Whitcomb, A. J. (1991). *I/O Psychology in Action*. Belmont, CA: Wadsworth Publishing Company.

Abdullah, A. (1992). *Understanding the Malaysian Workforce*. Kuala Lumpur: Malaysian Institute of Management.

Abrami, P., d'Appollonia, S., and Cohen, P. (1990). Validity of student ratings of instruction: what we know and what we do not know. *Journal of Educational Psychology*, 82, 219–231.

Adams, J. S. (1965). Inequity in social exchange. *Advances in Experimental Social Psychology*, 2, 267–300.

Adams, J. S., and Rosenbaum, W. B. (1962). The relationship of worker productivity to cognitive dissonance about wage inequities. *Journal of Applied Psychology*, 46, 161–164.

Adamson, P. (1993). Facts for Life: A Communication Challenge. New York: UNICEF, WHO, UNESCO and UNFPA.

Adjibolosoo, S. B. S. K. (ed.) (1995). *The Significance of the Human Factor in African Economic Development*. Westport, CT: Praeger.

Adler, N. J. (1986). *International Dimensions of Organizational Behavior*. Boston, MA: Kent Publishing.

Adler, N. J. (1993). An international perspective on the barriers to the advancement of women managers. *Applied Psychology: An International Review*, 42, 289–300.

Adorno, T. W., Frenkel-Brunswik, E., Levinson, D. J., and Sanford, R. N. (1950). *The Authoritarian Personality*. New York: Harper & Row.

Ager, A., Carr, S. C., MacLachlan, M., and Kaneka-Chilongo, B. (1996). Perceptions of tropical health risks in Mponda, Malawi: attributions of cause, suggested means of risk reduction and preferred treatment. *Psychology and Health*, 12, 23–31.

Ager, A. K. (1988). Planning sustainable services: principles for the effective targeting of resources in developed and developing nations. Paper presented at IASSMD World Congress, Dublin.

Ager, A. K. (1991). Effecting sustainable change in client behaviour: the role of the behavioural analysis of service environments. In B. Remington (ed.), *The Challenge of Severe Mental Handicap*. Chichester: Wiley.

Ager, A. K. (1996). Children, war, and psychological intervention. In S. C. Carr and J. F. Schumaker (eds), *Psychology and the Developing World* (pp. 162–172). Westport, CT: Praeger.

Airhihenbuwa, C. O. (1995). *Health and Culture: Beyond the Western Paradigm*. London: Sage.

Ajzen, I. (1991). The theory of planned behavior. *Organizational Behavior and Human Decision Processes, 50*, 179–211.

Alatas, S. H. (1978). *The Myth of the Lazy Native*. London: Frank Cass.

Ali, S., Nyirenda, T., and MacLachlan, M. (1994). The influence of traditional beliefs and practices on Chamba abuse in Malaŵi. Paper at World Congress of Social Psychiatry, Hamburg, Germany, 5–10 June.

Allen, K. R. (1987). An umbrella plan for expatriates. *Benefits and Compensation International, July*, 3–9.

Allen, T. (1992). Taking culture seriously. In T. Allen and A. Thomas (eds), *Poverty and Development in the 1990s* (pp. 331–346). Oxford: Oxford University Press.

Alvi, S. A., and Ahmed, S. W. (1987). Assessing organizational commitment in a developing country: Pakistan – a case study. *Human Relations, 40*, 267–280.

Anderson, N. R. (1992). Eight decades of employment selection interview research: a retrospective meta-review and prospective commentary. *European Work and Organizational Psychologist, 2*, 1–32.

Anderson, R. S., and Alexander, R. (1995). Innovate to grow. *Management: The Magazine of the Australian Institute of Management, October*, 8–10.

Anunobi, F. O. (1992) *The Implications of Conditionality, the International Monetary Fund and Africa: An Evaluation of IMF Policy Conditions in Africa*. University Press of America.

Ardila, R. (1996). Country profile: Colombia. *Psychology International, 7*, 5.

Argyle, M. (1989). *The Social Psychology of Work*. Harmondsworth: Penguin.

Aroni, S. (1995). Opening address. The Fourth International Symposium on the Role of Universities in Developing Areas/UNESCO Regional Seminar on Technology for Development, Melbourne, Australia, July.

Aronson, E., Blaney, N., Stephan, C., Sikes, J., and Snapp, M. (1978). *The Jigsaw Classroom*. Beverly Hills, CA: Sage.

Aronson, E., Willerman, B., and Floyd, J. (1966). The effects of a pratfall on increasing interpersonal attractiveness. *Psychonomic Science, 4*, 157–158.

Asch, S. E. (1956). Studies of independence and conformity: a minority of one against a unanimous majority. *Psychological Monographs: General and Applied, 70*, 1–70 [Whole No. 416].

Ashkanasy, N. M. (1994). Automatic categorisation and causal attribution: the effect of gender bias in supervisor responses to subordinate performance. *Australian Journal of Psychology, 46*, 177–182.

Assmar, E. M. L., and Rodriques, A. (1994). The value base of distributive justice: testing Deutsch's hypotheses in a different culture. *Revista interamericana de psicología, 28*, 1–11.

Augoustinos, M., and Walker, I. (1995). *Social Cognition: An Integrated Introduction*. London: Sage.

AusAID (Australian Agency for International development). (1995). Aid business, good business. Development Seminar, Regent Hotel, Sydney, April.

Austin, C. N. (1988). *Cross-Cultural Re-Entry: A Book of Readings*. Abilene, TX: Abilene Christian University.

Austin, W., and Walster, E. (1974). Reactions to confirmations and disconfirmations of expectations of equity and inequity. *Journal of Personality and Social Psychology*, 30, 208–216.

Axelrod, R. M. (1984). *The Evolution of Cooperation*. New York: Basic Books.

Azjen, I. (1985). From intentions to actions: a theory of planned behavior. In J. Kuhl and J. Beckman (eds), *Action-Control: From Cognition to Behavior* (pp. 11–39). Heidelberg: Springer.

Baddeley, A., Gardner, J. M., and Grantham-McGregor, S. (1995). Cross-cultural cognition: developing tests for developing countries. *Applied Cognitive Psychology*, 9, 173–195.

Bagozzi, R. P. (1981). Attitudes, intentions, and behavior: a test of some key hypotheses. *Journal of Personality and Social Psychology*, 41, 607–627.

Baillie, A., and Porter, N. (1996). Encouraging a deep approach to the undergraduate psychology curriculum: an example and some lessons from a third year course. *Psychology Teaching Review*, 5, 14–24.

Baker, W. K. (1995). Allen and Meyer's 1990 longitudinal study: a reanalysis and reinterpretation using structural modeling. *Human Relations*, 48, 169–186.

Baldwin, S., Magjuka, H., and Loher, E. (1991). The perils of participation: effects of choice of training on trainee motivation and learning. *Personnel Psychology*, 44, 51–65.

Bales, R. F. (1955). How people interact in conferences. *Scientific American*, 192, 31–35.

Ballard, B. (1987). Academic adjustment: the other side of the export dollar. *Higher Education Research and Development*, 6, 109–119.

Barbachon, G. (1968). The diffusion of scientific and technical knowledge. *Journal of Social Issues*, 43 79–84.

Barker, M., Child, C., Gallois, C., Jones, E., and Callan, V. J. (1991). Difficulties of overseas students in social and academic situations. *Australian Journal of Psychology*, 43, 79–84.

Bartlett, F. C. (1995). *Remembering: A Study in Experimental and Social Psychology*. Cambridge: Cambridge University Press.

Bau, L., and Dyck, M. (1992). Predicting the peacetime performance of military officers: officer selection in the Papua New Guinea Defence Force. *South Pacific Journal of Psychology*, 5, 27–37.

Bazar, J. (1994). Vietnamese psychologists meet challenges of a changing society. *Psychology International*, 5, 1–3.

Beacham, R. H. S. (1979). *Pay Systems: Principles and Techniques*. London: Heinemann.

Beaman, A., Barnes, P. J., Klentz, B., and McQuirk, B. (1978). Increasing helping rates through information dissemination: teaching pays. *Personality and Social Psychology Bulletin*, 4, 406–411.

Beardsley, T. (1993). Never give a sucker an even break. *Scientific American*, October, 12.

Bedeian, A. G. (1995). Workplace envy. *Organizational Dynamics*, 23, 49–56.

Bejar Navarro, R. (1986). *El Méxicano*. Mexico City: Universidad Nacional Autónoma de México.

Belbin, R. M. (1981). *Management Teams: Why They Succeed or Fail*. New York: Wiley.

Bem, D. J. (1967). Self-perception: an alternative interpretation of cognitive dissonance phenomena. *Psychological Review, 74*, 183–200.

Bem, D. J. (1972). Self-perception theory. *Advances in Experimental Social Psychology, 6*, 1–62.

Berg, E. (1993). *Rethinking Technical Cooperation. Reforms for Capacity Building in Africa*. New York: Regional Bureau for Africa, United Nations Development Programme and Development Alternatives, Inc.

Berkowitz, L., Fraser, C., Treasure, F. P., and Cochran, S. (1987). Pay, equity, job qualifications, and comparisons in pay satisfaction. *Journal of Applied Psychology, 72*, 544–551.

Berman, J. J., Murphy-Berman, U., and Singh, P. (1985). Cross-cultural similarities and differences in perceptions of fairness. *Journal of Cross-cultural Psychology, 16*, 55–67.

Berne, E. (1964). *Games People Play*. New York: Grove Press.

Berrenberg, J. L., Rosnik, D., and Kravcisin, N. J. (1991). Blaming the victim: when disease prevention programs misfire. *Current Psychology Research and Reviews, 9*, 415–420.

Berry, J. (1979). A cultural psychology of social behavior. *Advances in Experimental Social Psychology, 12*, 127–207.

Berry, J. W., Kalin, R., and Taylor, D. M. (1977). *Multiculturalism and ethnic attitudes in Canada*. Ottawa: Minister of Supply and Services.

Bersheid, E., and Walster, E. (1967). When does a harmdoer compensate a victim? *Journal of Personality and Social Psychology, 6*, 435–441.

Bhattacharya, S. P. S. P. (1994). The application of transactional analysis in a participatory forest management program. *Transactional Analysis Journal, 24*, 286–290.

Bierhoff, H. W., and Klein, R. (1990). Prosocial behaviour. In M. Hewstone, W. Stroebe, J. P. Codol, and G. M. Stephenson (eds), *Introduction to Social Psychology* (pp. 246–263). Oxford: Basil Blackwell.

Bishop, G. (1996). East meets West: illness, cognition and behaviour in Singapore. 10th European Health Psychology Society Conference, Dublin, 4–6 September.

Blake, R. R., and Mouton, J. S. (1978). *The New Managerial Grid*. Houston, TX: Gulf Publishing Company.

Blunt, P. (1983). *Organizational Theory and Behaviour: An African Perspective*. New York: Longman.

Blunt, P., and Jones, M. L. (1986). Managerial motivation in Kenya and Malawi: a cross-cultural comparison. *Journal of Modern African Studies, 24*, 165–175.

Blunt, P., and Jones, M. L. (1992). *Managing Organisations in Africa*. Berlin: De Gruyter.

Bochner, S., and Wicks, P. (eds) (1972). *Overseas Students in Australia*. Sydney: New South Wales University Press.

Bond, R., and Smith, P. B. (1996). Culture and conformity: a meta-analysis of studies using Asch's (1952b, 1956) line judgment task. *Psychological Bulletin, 119*, 111–137.

Booker-Weiner, J. (1995). Building business and management programs in the post-Communist world. In S. Aroni, and T. Adams (eds), *Proceedings Vol 1: The Fourth International Symposium on the Role of Developing Countries in Developing Areas* (pp. 1–15). Melbourne, Australia: UNESCO/INRUDA/IDTC/RMIT.

Bossert, T. J. (1990). Can they get along without us? Sustainability of donor-supported health projects in Central America and Africa. *Social Sciences and Medicine, 30*, 1015–1023.

Boucebci, M., and Bensmail, B. (1982). Psychopathological aspects of decompensations observed among volunteer workers abroad. *Annales medico psychologiques, 140*, 677–680.

Bourhis, R. Y., Giles, H., and Tajfel, H. (1973). Language and determinant of Welsh identity. *European Journal of Social Psychology, 13*, 321–350.

Bowa, M., and MacLachlan, M. (1994). No congratulations in Chichewa: deterring achievement motivation in Malawi. In S. S. Chiotha (ed.), *Research and Development III*. Zomba: University of Malawi.

Braybrooke, D., and Lindblom, C. E. (1963). *A Strategy of Decision*. New York: Free Press.

Brehm, J. W. (1966). *A Theory of Psychological Reactance*. New York: Academic Press.

Brehm, J. W. (1972). *Responses to Loss of Freedom: A Theory of Psychological Reactance*. Morristown, NJ: General Learning Press.

Brett, E. A. (1996). The participatory principle in development projects: the costs and benefits of co-operation. *Public Administration and Development, 16* (1), 5–19.

Brewer, A. M. (1995). *Change Management: Strategies for Australian Organisations*. Sydney: Allen & Unwin.

Brinkerhoff, D. W., and Goldsmith, A. A. (1992). Promoting the sustainability of development institutions: a framework for strategy. *World Development, 20*, 369–383.

Brislin, R. W., Cushner, K., Cherrie, C., and Yong, M. (1986). *Intercultural Interactions: A Practical Guide*. Beverly Hills, CA: Sage.

Brockner, J., and Adsit, L. (1986). The moderating impact of sex on the equity-satisfaction relationship: a field study. *Journal of Applied Psychology, 71*, 585–590.

Brown, M., and Ralph, S. (1996). Barriers to women managers' advancement in education in Uganda. *International Journal of Educational Management, 10*, 18–23.

Brunning, H., Cole, C. and Huffington, C. (1990). *The Change Directory: Key Issues in the Organisation of Change and the Management of Change*. Leicester: British Psychological Society (Division of Clinical Psychology).

Bruntland, G. H. (1987). What is sustainable development? In Panos Institute (eds), *Towards Sustainable Development* (pp. viii–x). London: Panos Institute.

Burgoyne, C. B., and Routh, D. A. (1991). Constraints on the use of money as a gift at Christmas: the role of status and intimacy. *Journal of Economic Psychology, 12*, 47–69.

Burnkrant, R. E., and Page, T. J. (1982). An examination of the convergent, discriminate, and predictive validity of Fishbein's behavioral intention model. *Journal of Marketing Research, 19*, 550–561.

Buzzotta, V. R., Lefton, R. E., and Sherberg, M. (1972). *Effective Selling through Psychology*. New York: Wiley.

Byrne, D. (1971). *The Attraction Paradigm*. New York: Academic Press.

213

Callahan, M. R. (1989). Preparing the new global manager. *Training and Development Journal, March*, 29–32.

Cantril, J. G. (1991). Inducing health care voluntarism through sequential requests: perceptions of effort and novelty. *Health Communication, 3*, 59–74.

Cantril, J. G., and Seibold, D. R. (1986). The perceptual contrast explanation of sequential request strategy effectiveness. *Human Communication Research, 13*, 253–267.

Carbert, C., and Chikarovski, K. (1997). *Remuneration Systems, Trends and Developments.* Sydney: College of Organisational Psychologists.

Carnevale, P. J. (1995). Property, negotiation, and culture. In R. M. Kramer and D. M. Messick (eds), *Negotiation as a Social Process* (pp. 309–323). Thousand Oaks, CA: Sage.

Carnevale, P. J., and Pruitt, D. G. (1992). Negotiation and mediation. *Annual Review of Psychology, 43*, 531–582.

Carr, S. C. (1993). The family in Malaŵi as a resource for HIV/AIDS prevention. *Community Alternatives: International Journal of Family Care, 5*, 123–125.

Carr, S. C. (1994a). Developing psychology in Malaŵi: one step on from the SWOT. *Psychology Teaching Review, 3*, 1–6.

Carr, S. C. (1994b). Generating the velocity for overcoming motivational gravity in LDC business organizations. *Journal of Transnational Management Development, 1*, 33–56.

Carr, S. C. (1996b). Social psychology and the management of aid. In S. C. Carr and J. F. Schumaker (eds), *Psychology and the Developing World* (pp. 103–118). Westport, CT: Praeger.

Carr, S. C. (1996c). Social psychology in Malaŵi: 'historical' or 'developmental'? *Psychology and Developing Societies, 8*, 177–197.

Carr, S. C., Ager, A., Nyando, C., Moyo, K., Titeca, A., and Wilkinson, M. (1994). A comparison of chamba (marijuana) abusers and general psychiatric admissions in Malaŵi. *Social Science and Medicine, 39*, 401–406.

Carr, S. C., Chipande, R., and MacLachlan, M. (1998). Expatriate aid salaries in Malaŵi: a doubly demotivating influence? *International Journal of Educational Development, 18*.

Carr, S. C., Ehiobuche, E., Rugimbana, R., and Munro, D. (1996). Expatriates' ethnicity and their effectiveness: 'similarity-attraction' or 'inverse resonance?' *Psychology and Developing Societies, 8*, 265–282.

Carr, S. C., and MacLachlan, M. (1993). Asserting psychology in Malaŵi. *The Psychologist, 6*, 408–413.

Carr, S. C., and MacLachlan, M. (1996a). Towards a Malaŵian psychology. *Journal of Psychology in Africa, 6*, 100–119.

Carr, S. C. and MacLachlan, M. (1996b). Managing Tropical Health: psychology for development? *British Medical Anthropology Review, 2*, 41–47.

Carr, S. C., and MacLachlan, M. (1997). Motivational gravity and organizational culture. In D. Munro, J. F. Schumaker, and S. C. Carr (eds), *Motivation and Culture.* New York: Routledge.

Carr, S. C., and MacLachlan, M. (in press). Actors, observers, and attributions for Third World poverty: contrasting perspectives from Malaŵi and Australia. *Journal of Social Psychology.*

Carr, S. C., MacLachlan, M., and Campbell, D. (1995). Psychological research for development: towards tertiary collaboration. In S. Aroni, and T. Adams (eds), *Proceedings Vol. 1 – The Fourth International Symposium on the Role of Universities in Developing Areas* (pp. 85–105). Melbourne, Australia: UNESCO/INRUDA/IDTC/RMIT.

Carr, S. C., MacLachlan, M., Heathcote, A., and Heath, R. A. (1997). Developing psychology in Malaŵi: a lesson for educational testing? *Psychology Teaching Review*, 6, 157–169.

Carr, S. C., MacLachlan, M., Kachedwa, M., and Kanyangale, M. (1997). The meaning of work in Malaŵi. *Journal of International Development*, 9, 899–911.

Carr, S. C., MacLachlan, M., and Schultz, R. (1995). Pacific Asia psychology: ideas for development? *South Pacific Journal of Psychology*, 8, 1–18.

Carr, S. C., MacLachlan, M., Zimba, C., and Bowa, M. (1995a). Managing motivational gravity in Malawi. *Journal of Social Psychology*, 135, 659–662.

Carr, S. C., MacLachlan, M., Zimba, C., and Bowa, M. (1995b). Community aid abroad: a Malaŵian perspective. *Journal of Social Psychology*, 135, 781–783.

Carr, S. C., and Mansell, D. (1995). *Higher Education Development Project: Report on Joint Selection*. Melbourne: International Development Technologies Centre.

Carr, S. C., McKay, D., and Rugimbana, R. (1997). *International Students and a Changing Educational System: Why Aren't we Applying Psychology?* Newcastle, Australia: University of Newcastle.

Carr, S. C., McLoughlin, D., Hodgson, M., and MacLachlan, M. (1996). Effects of unreasonable pay discrepancies for under- and overpayment on double demotivation. *Genetic, Social, and General Psychology Monographs*, 122, 477–494.

Carr, S. C., and Munro, D. (1994). A new style of psychology for development studies. *Development Bulletin*, 32, 60–63.

Carr, S. C., Munro, D., and Bishop, G. D. (1996). Attitude assessment in non-Western countries: critical modifications to Likert scaling. *Psychologia: An International Journal of Psychology in the Orient*, 39, 55–59.

Carr, S. C., Pearson, S. A., and Provost, S. (1996). Learning to manage motivational gravity: An application for group polarization. *Journal of Social Psychology*, 136, pp. 251–4.

Carr, S. C., Rugimbana, R., and Walkom, E. (1997). *Inverse Resonance and Aid Selection: Some Preliminary Support from Tanzania*. Newcastle, Australia: University of Newcastle.

Carr, S. C., Watters, P., and MacLachlan, M. (1996). Beyond cognitive tolerance: towards the edge of chaos? 10th European Health Psychology Society Conference, Dublin, Ireland, 4–6 September.

Cassen, R. (1994). *Does Aid Work?* Oxford: Oxford University Press.

Castro, M. A. (1974). Reactions to receiving aid as a function of cost to donor and opportunity to aid. *Journal of Applied Social Psychology*, 4, 194–209.

Cederblad, M., and Rahim, S. (1986). Effects of rapid urbanization on child behaviour and health in a part of Khartoum, Sudan: socio–economic changes 1965–1980. *Social Science and Medicine*, 22, 713–721.

Chan, C., Law, C. K., and Kwok, R. (1992). Attitudes of women toward work in socialist and capitalist cities: a comparative study of Beijing, Guangzhou, and Hong Kong. *Canadian Journal of Community Mental Health*, 11, 187–200.

Chan, D. K., Triandis, H. C., Carnevale, P. J., Tam, A., and Bond, M. H. (1994).

Comparing Negotiation across Cultures: Effects of Collectivism, Relationship between Negotiators, and Concession Pattern on Negotiation Behavior. Urbana, IL: University of Illinois.

Charng, H. W., Piliavin, J. A., and Callero, P. L. (1988). Role identity and reasoned action in the prediction of repeated behavior. *Social Psychology Quarterly, 51,* 303–317.

Cherniss, C. (1980). *Staff Burnout: Job Stress in the Human Services.* Beverly Hills, CA: Sage.

Chidgey, J. E., and Carr, S. C. (1996). Relevance of the sporting metaphor for overcoming motivational gravity in Australia. 25th Meeting of Australasian Social Psychologists, Canberra, May.

Chimombo, M., and MacLachlan, M. (1995). *The AIDS Challenge.* Lilongwe: UNICEF (Malaŵi).

Chinese Culture Connection (1987). Chinese values and the search for culture-free dimensions of culture. *Journal of Cross-Cultural Psychology, 18,* 143–164.

Cialdini, R. B., Vincent, J. E., Lewis, S. K., Catalan, J., Wheeler, D., and Darby, B. L. (1975). A reciprocal concessions procedure for inducing compliance: The door-in-the-face technique. *Journal of Personality and Social Psychology, 21,* 206–215.

Clark, A. W., and McCabe, S. (1970). Leadership beliefs of Australian Managers. *Journal of Applied Psychology, 54,* 1–6.

Clarke, R. (1979). *The Japanese Company.* New Haven, CT: Yale University Press.

Coch, L., and French, J. R. P. (1948). Overcoming resistance to change. *Human Relations, 1,* 512–532.

Cohen, A. R. (1962). An experiment on small rewards for discrepant compliance and attitude change. In J. W. Brehm and A. R. Cohen (eds), *Explorations in Cognitive Dissonance* (pp. 73–78). New York: Wiley.

Collier, J., and Burke, A. (1986). Racial and sexual discrimination in the selection of students for London medical schools. *Medical Education, 20,* 86–90.

Colling, T. (1992). *Beyond Mateship: Understanding Australian Men.* Sydney: Simon & Schuster.

Commission of the European Communities. (1977). *The perception of poverty in Europe.* Brussels: EEC.

Conway, R. (1971). *The Great Australian Stupor.* Melbourne, Australia: Sun Books Pty.

Copestake, J. (1996). NGO–state collaboration and the new policy agenda – the case of subsidized credit. *Public Administration and Development, 16* (1), (Feb.) 21–30.

Costa, P. T., and McCrae, R. R. (1988). From catalog to classification: Murray's needs and the five factor model. *Journal of Personality and Social Psychology, 55,* 258–265.

Cottam, M. L. (1989). Cognitive psychology and bargaining behavior: Peru versus the MNCs. *Political Psychology, 10,* 445–475.

Cox, E. (1995). *A Truly Civil Society: The 1995 Charles Boyer Lectures.* Sydney: Australian Broadcasting Corporation (ABC) Books.

Craig, D., and Porter, D. (1994). The Phoenix of the subject in development. *Why Psychology? Australian Psychological Society Annual Conference Abstracts, 29,* 58.

Crano, W. D., and Sivacek, J. (1984). The influence of incentive-aroused ambivalence on overjustification effects in attitude change. *Journal of Experimental Social Psychology*, 20, 137–158.

Daroesman, I. P., and Daroesman, R. (1992). *Degrees of Success: A Tracer Study of Australian Government Sponsored Indonesian Fellowships 1970–1989*. Canberra: AusAID (Australian Agency for International Development) and IDP (International Development Program) Education Australia Ltd.

Daun, A. (1991). Individualism and collectivity among Swedes. *Ethnos*, 56, 165–172.

David, B., and Turner, J. C. (1996). Studies in self-categorization and minority conversion: is being a member of the outgroup an advantage? *British Journal of Social Psychology*, 35, 179–199.

Dawkins, J. (1993). *Australia and the World Bank 1992–3*. Canberra: Australian Government Publishing Service.

Deaux, K., and Wrightsman, L. S. (1984). *Social Psychology in the 80s*. Monterey, CA: Brooks/Cole Publishing Company.

Deci, E. L. (1975). *Intrinsic Motivation*. New York: Plenum Publishing.

Deci, E. L. (1987). *The Psychology of Self Determination*. New York: Lexington Books.

Deci, E. L., and Ryan, R. M. (1985). *Intrinsic Motivation and Self Determination in Human Behavior*. New York: Plenum.

DEET (Department of Employment, Education, and Training) (1991). *Country Education Profiles: Indonesia*. Canberra: Australian Government Publishing Service.

De Jong, W. (1979). An examination of self-perception mediation of the foot-in-the-door effect. *Journal of Personality and Social Psychology*, 37, 2221–2239.

Denys, L. O. (1971). Expatriates and cross-cultural communication. *Continuous Learning*, 10, 53–60.

De Paulo, B. M., Brown, P. L., Ishii, S., and Fisher, J. D. (1981). Help that works: the effects of aid on subsequent task performance. *Journal of Personality and Social Psychology*, 41, 478–487.

Dillard, J. P., Hunter, J. E., and Burgoon, M. (1984). Sequential request strategies: Meta-analysis of foot-in-the-door and door-in-the-face. *Human Communication Research*, 10, 461–488.

Doise, W. (1969). Intergroup relations and polarization of individual and collective judgements. *Journal of Personality and Social Psychology*, 12, 136–143.

Dooley, B. (1996). At work away from work. *The Psychologist*, 9, 155–158.

Dore, R. (1994). Why visiting sociologists fail. *World Development*, 22, 1425–1436.

Dorward, D. (1996). Africa and development in the 21st century. *Development Bulletin*, 37, 4–7.

Downing, J., and Harrison, T. C. (1990). Dropout prevention: a practical approach. *School Counselor*, 38, 67–74.

Drucker, P. F. (1954). *The Practice of Management*. New York: Harper & Row.

Drucker, P. F. (1988). *Management*. London: Heinemann.

Dubbey, J. M., Chipofoya, C. C., Kandawire, J. A. K., Kasomekera, Z. M., Kathamalo, O. J., and Machili, G. G. (1991). How effective is our university? A study of graduates from the University of Malaŵi. *Higher Education Quarterly*, 45, 219–233.

Dunphy, D., and Stace, D. (1993). The strategic management of corporate change. *Human Relations*, 46, 905–920.

Earley, P. C. (1993). East meets West meets mid East: further explorations of collectivistic and individualistic work groups. *Academy of Management Journal,* *36*, 319–348.

Eayrs, C. B., and Ellis, N. (1990). Charity advertising: For or against people with a mental handicap. *British Journal of Social Psychology, 29,* 349–360.

Edwards, M. and Hulme, D. (1994) NGOs and development: performance and accountability in the 'new world order'. Position paper for an eponymous conference held at Manchester University and sponsored by Save the Children Fund.

Ellerman, D. A., and Smith, E. P. (1983). Generalised and individual bias in the evaluation of the work of women: sexism in Australia. *Australian Journal of Psychology, 35,* 71–79.

Elliot, E., Pirrs, M., and McMaster, J. (1992). Nurses' views of parasuicide in a developing country. *The International Journal of Social Psychiatry, 38,* 273–279.

Emilia, O., and Mulholland, H. (1991). Approaches to learning of students in an Indonesian medical school. *Medical Education, 25,* 462–470.

Engle, P. (1996). Combating malnutrition in the developing world. In S. C. Carr and J. F. Schumaker (eds), *Psychology and the Developing World* (pp. 153–161). Westport, CT: Praeger.

Evans-Pritchard, E. E. (1976). *Witchcraft, oracles, and magic among the Azande.* Oxford, UK: Clarendon Press.

Eze, N. (1985). Sources of motivation among Nigerian managers. *Journal of Social Psychology, 125,* 341–345.

Fanon, F. (1985). *The Wretched of the Earth.* Harmondsworth: Penguin.

Fanon, F. (1991). *Black Skin, White Masks.* London: Pluto Press.

Feagin, J. R. (1972). Poverty: We still believe that God helps those who help themselves. *Psychology Today, 6,* 101–129.

Feather, N. T. (1974). Explanations of poverty in Australian and American samples: The person, society or fate? *Australian Journal of Psychology, 26,* 199–216.

Feather, N. T. (1994). Human values and their relation to justice. *Journal of Social Issues, 50,* 129–151.

Feather, N. T. (1994). Attitudes toward high achievers and reactions to their fall: theory and research concerning tall poppies. *Advances in Experimental Social Psychology, 26,* 1–73.

Feather, N. T., and McKee, I. R. (1993). Global self-esteem and attitudes toward the high achiever for Australian and Japanese students. *Social Psychology Quarterly, 56,* 65–76.

Festinger, L. (1950). Informal social communication. *Psychological Review, 57,* 271–282.

Festinger, L. (1954). A theory of social comparison processes. *Human Relations, 1,* 117–140.

Festinger, L. (1957). *A Theory of Cognitive Dissonance.* Stanford, CA: Stanford University Press.

Fiedler, F. E. (1978). The contingency model and the dynamics of the leadership process. *Advances in Experimental Social Psychology, 11,* 59–112.

Fijneman, Y. A., Willemsen, M. E., and Poortinga, Y. H. (1996). Individualism–collectivism: An empirical study of a conceptual issue. *Journal of Cross-cultural Psychology, 27,* 381–402.

Fishbein, M., and Ajzen, I. (1975). *Belief, Attitude, Intention, and Behavior: An Introduction to Theory and Research.* Reading, MA: Addison-Wesley.

Fisher, J. D., de Paulo, B. M., and Nadler, A. (1981). Extending altruism beyond the atruistic act: the mixed effects of aid to the help recipient. In J. P. Rushton and R. M. Sorrentino (eds), *Altruism and Helping Behavior* (pp. 367–422). Hillsdale, NJ: Erlbaum.

Flanagan, J. C. (1954). The critical incident technique. *Psychological Bulletin*, *51*, 327–349.

Florkowski, G. W., and Schuster, M. H. (1992). Support for profit sharing and organizational commitment: a path analysis. *Human Relations*, *45*, 507–523.

Forgas, J. P. (1986). *Interpersonal Relations*. Sydney: Pergamon Press.

Forgas, J. P. (1986). *Interpersonal Behaviour: The Psychology of Social Interaction*. Sydney: Pergamon Press.

Foster, J. J., Moss, B., and Wilkie, D. (1996). Selecting psychology staff. *The Psychologist*, *9*, 364–367.

Fountain, S. (1995). *Education for Development*. London: Hodder & Stoughton/ UNICEF.

Fowler, A. (1996). Demonstrating NEO performance: problems and possibilities. *Development in Practice*, *6*, 58–65.

Fox, C. (1994). Educational development cooperation, ethics, and the role of the consultant. *Development Bulletin*, *30*, 43–45.

Freed, J. E., and Burack, E. H. (1996). Employee involvement and TQM. *Organizational Development Journal*, *14*, 19–29.

Freedman, J. L., and Fraser, S. C. (1966). Compliance without pressure: The foot-in-the-door technique. *Journal of Personality and Social Psychology*, *4*, 195–202.

Freire, P. (1972). *Pedagogy of the Oppressed*. Harmondsworth: Penguin.

French, J. R. P., Jr. and Raven, B. H. (1959). The bases of social power. In D. Cartwright (ed.), *Studies in Social Power*. Ann Arbor, MI: University of Michigan Press.

Friend, R. M., and Neale, J. M. (1972). Children's perceptions of success and failure: An attributional analysis of the effects of race and social class. *Developmental Psychology*, *7*, 124–128.

Friere, P. (1970). *Pedagogy of the Oppressed*. New York: Seabury Press.

Frisch, M. B., and Gerrard, M. (1981). Natural helping systems: a survey of Red Cross volunteers. *American Journal of Community Psychology*, *9*, 567–579.

Fromkin, H. L. (1972). Feelings of interpersonal undistinctiveness: an unpleasant affective state. *Journal of Experimental Research in Personality*, *6*, 178–185.

Fry, P. S., and Ghosh, R. (1980). Attributions of success and failure: Comparison of cultural differences between Asian and Caucasian children. *Journal of Cross-Cultural Psychology*, *11*, 343–363.

Furnham, A. (1982a). Explaining poverty in India: A study of religious group differences. *Psychologia: An International Journal of Psychology in the Orient*, *25*, 236–243.

Furnham, A. (1982b). The perception of poverty among adolescents. *Journal of Adolescence*, *5*, 135–147.

Furnham, A. (1983). Attributions for affluence. *Personality and Individual Differences*, *4*, 31–40.

Furnham, A. (1993). Just world beliefs in twelve societies. *Journal of Social Psychology*, *133*, 317–329.

219

Furnham, A., and Gunter, B. (1984). Just world beliefs and attitudes towards the poor. *British Journal of Social Psychology*, 23, 265–269.

Gallin, R. S. (1989). Women and work in rural Taiwan: building a contextual model linking employment and health. *Journal of Health and Social Behavior*, 30, 374–385.

Gallois, C., Callan, V. J., and Palmer, J. A. M. (1992). The influence of applicant communication style and interviewer characteristics on hiring decisions. *Journal of Applied Social Psychology*, 22, 1041–1060.

Gallup, G. H. (1972). *The Gallup Poll: Public Opinion 1935–1971*. New York: Random House.

Gergen, K. and Gergen, M. (1981). *Social Psychology*. New York: Harcourt Brace Jovanovich.

Gergen, K. J. (1973). Social psychology as history. *Journal of Personality and Social Psychology*, 26, 309–320.

Gergen, K. J. (1974). Toward a psychology of receiving help. *Journal of Applied Social Psychology*, 4, 187–193.

Gergen, K. J. (1994). *Toward Transformation in Social Knowledge*. London: Sage.

Gergen, K. J., Ellsworth, P., Maslach, C., and Seipel, M. (1975). Obligation, donor resources, and reactions to aid in three cultures. *Journal of Personality and Social Psychology*, 31, 390–400.

Gergen, K. J., and Gergen, M. M. (1971). International assistance from a psychological perspective. In G. W. Keeton and G. Schwarzenberger (eds), *Year Book of World Affairs 1971* (pp. 87–103). London: Stevens & Sons.

Gergen, K. J., and Gergen, M. M. (1974). Understanding foreign assistance through public opinion. In G. W. Keeton and G. Schwarzenberger (eds), *Year Book of World Affairs 1974* (pp. 125–140). London: Stevens & Sons.

German, T., and Randel, J. (1995). *The Reality of Aid: An Independent Review of International Aid*. London: Earthscan.

Gertzel, C. (1994). *The New World Order: Implications for Development*. Canberra: Australian Development Studies Network, Australian National University, Briefing Paper No. 35.

Gertzel, C. (1996). Towards a better understanding of poverty in Africa in the late twentieth century. *Australian Development Studies Network*, 42, 1–6.

Giacalone, R. A., and Beard, J. W. (1994). Impression management, diversity, and international management. *American Behavioral Scientist*, 37, 621–636.

Gibb, C. A. (1969). Leadership. In G. Lindzey and E. Aronson (eds), *Handbook of Social Psychology Vol. IV* (pp. 205–282). Reading, MA: Addison-Wesley.

Gleitman, H. (1986). *Psychology*. New York: Norton.

Godin, G. *et al.* (1996). Cross-cultural testing of three social cognitive theories: an application to condom use. *Journal of Applied Social Psychology*, 26, 1556–1586.

Godwin, N. (1994). A distorted view: myths and images of developing countries. *Development Bulletin*, 30, 46–48.

Goffman, E. (1959). *The Presentation of Self in Everyday Life*. New York: Doubleday.

Goldman, M. (1986). Compliance employing a combined foot-in–the-door and door-in-the-face procedure. *Journal of Social Psychology*, 126, 111–116.

Gologor, E. (1977). Group polarization in a non-risk-taking culture. *Journal of Cross-Cultural Psychology*, 8, 331–346.

Good, C. M. (1987). *Ethnomedical systems in Africa: Patterns of Traditional Medicine in Rural and Urban Kenya*. London: The Guildford Press.

Goold, S. (1995). Why are there so few Aboriginal registered nurses? In G. Grey and R. Pratt (eds), *Issues in Australian Nursing Vol. 4* (pp. 235–252). Melbourne: Churchill-Livingstone.

Gopal, M. (1995). Teaching psychology in a Third World setting. *Psychology and Developing Societies, 7,* 21–45.

Gouldner, A. W. (1960). The norm of reciprocity: a preliminary statement. *American Sociological Review, 25,* 161–178.

Gow, D. D. (1991). Collaboration in development consulting: stooges, hired guns, or musketeers? *Human Organization, 50,* 1–15.

Grace, C. R., Bell, P. A., and Sugar, J. (1988). Effects of compliance on spontaneous and asked-for helping. *Journal of Social Psychology, 128,* 525–532.

Greenberg, J. (1990). Employee theft as a reaction to underpayment inequity: the hidden costs of pay cuts. *Journal of Applied Psychology, 73,* 606–613.

Greenberg, J., and Ornstein, S. (1983). High status job title as compensation for underpayment: a test of equity theory. *Journal of Applied Psychology, 68,* 285–297.

Greenwood, G. (1974). *Approaches to Asia: Australian Postwar Policies and Attitudes*. Sydney: McGraw-Hill.

Gregory, C. J., Higginbotham, N., and Shea, J. D. (1997). Multi-component solar protection interventions: the impact of modelling and 'trickle down'? *South Pacific Journal of Psychology, 9.*

Griffin, W., and Oheneba-Sakyi, Y. (1993). Sociodemographic and political correlates of university students' causal attributions for poverty. *Psychological Reports, 73,* 795–800.

Grill, B. (1995). Afrique. *GEO: Un nouveau monde la Terre, 202 (December),* 48–76.

Guimond, S., and Palmer, D. L. (1990). Type of academic training and causal attributions for social problems. *European Journal of Social Psychology, 20,* 61–75.

Gurstein, M., and Klee, J. (1996). Towards a management renewal of the United Nations – Part I. *Public Administration and Development, 16,* 43–56.

Guth, W., Schmittberger, R., and Schwarze, B. (1982). An experimental analysis of ultimatum bargaining. *Journal of Economic Behavior and Organization, 3,* 367–388.

Haley, A., and X., Malcolm. (1987). *The Autobiography of Malcolm X*. Harmondsworth: Penguin.

Halpern, D., and Osofsky, S. (1990). A dissenting view of MbO. *Personnel Management, 19,* 321–330.

Handy, C. B. (1985). *Understanding Organizations*. Harmondsworth: Penguin.

Harpaz, I. (1989). Non-financial employment commitment: a cross-national comparison. *Journal of Occupational Psychology, 62,* 147–150.

Harper, D. J., and Manasse, P. R. (1992). The Just World and the Third World: British explanations for poverty abroad. *Journal of Social Psychology, 132,* 783–785.

Harper, D. J., Wagstaff, G. F., Newton, J. T., and Harrison, K. R. (1990). Lay causal perceptions of Third World poverty and the just world theory. *Social Behavior and Personality, 18,* 235–238.

Harré, R., and Van Langenhove, L. (1996). Varieties of positioning. *Theory for the Theory of Social Behaviour, 21,* 393–407.

Harrison, J. K., Chadwick, M., and Scales, M. (1996). The relation between cross-cultural adjustment and the personality variables of self-efficacy and self-monitoring. *International Journal of Intercultural Relations, 20*, 167–188.

Harrison, J. R., and Bazerman, M. H. (1995). Regression to the mean, expectation inflation, and the winner's curse in organizational contexts. In R. M. Kramer and D. M. Messick (eds), *Negotiation as a Social Process* (pp. 69–94). Thousand Oaks, CA: Sage.

Harriss, J. (1995). 'Japanization': context and culture in the Indonesian automotive industry. *World Development, 23*, 117–128.

Harrod, J. (1974). Problems of the United Nations specialised agencies at the quarter century. In G. W. Keeton and G. Schwarzenberger (eds), *The Year Book of World Affairs 1974* (pp. 187–203). London: Stevens & Sons.

Hartman, S. J., Fok, L. Y., and Villere, M. F. (1997). The impact of culture on the relationship between organizational citizenship behavior and equity sensitivity. *Journal of Transnational Management Development, 2*, 33–50.

Heider, F. (1958). *The Psychology of Interpersonal Relations*. New York: Wiley.

Hernandez Alvarez, O. (1994). 'Latin America.' In A. Trebilcock (ed.), *Towards Social Dialogue*. Geneva: ILO.

Herriot, P. (1991). The selection interview. In P. Warr (ed.), *Psychology at Work* (pp. 139–159). Harmondsworth: Penguin.

Herzberg, F. (1966). *Work and the Nature of Man*. Cleveland, OH: World Publishing.

Hesketh, B., and Gardner, D. (1993). Person–environment fit models: A reconceptualization and empirical test. *Journal of Vocational Behavior, 42*, 315–337.

Hofstede, G. (1980a). *Culture's Consequences: International Differences in Work Related Values*. Beverly Hills, CA: Sage.

Hofstede, G. (1980b). Motivation, leadership, and organization: do American theories apply abroad? *Organizational Dynamics, Summer*, 42–63.

Hofstede, G. (1984). The cultural relativity of the Quality of Life concept. *Academy of Management Review, 9*, 389–398.

Hofstede, G. (1985). Cultural differences in teaching and learning. Colloquium on Selected Issues in International Business, Honolulu, HA, August.

Hofstede, G. (1991). *Culture and Organizations: Software of the Mind*. Beverly Hills, CA: Sage.

Hofstede, G., and Bond, M. H. (1988). The Confucius connection: from cultural roots to economic growth. *Organizational Dynamics, 16*, 4–21.

Hogg, M. A., Turner, J. C. and Davidson, B. (1990). Polarized norms and social frames of reference: a test of the self-categorization theory of group polarization. *Basic and Applied Psychology, 11*, 77–100.

Hollander, E. P. (1964). *Leaders, Groups, and Influence*. New York: Oxford University Press.

Hope, A., and Timmel, S. (1995). *Training for Transformation, Vol. 1*. Gweru, Zimbabwe: Mambo Press.

House, W., and Zimalirana, G. (1992). Rapid population growth and its implications for Malawi. *Malawi Medical Journal, 8*, 46–64.

Howarth, I., and Croudace, T. (1995). Improving the quality of teaching in universities: a problem for occupational psychologists? *Psychology Teaching Review, 4*, 1–11.

Hughes, P. (1988). Aboriginal culture and learning style: a challenge for academics in higher educational institutions. University of New England, NSW, Australia. Frank Archibald Memorial Lecture.

Hui, C. H. (1990). Work attitudes, leadership styles, and managerial behaviors in different cultures. In R. W. Brislin (ed.), *Applied Cross-cultural Psychology* (pp. 186–208). Newbury Park, CA: Sage.

Hui, C. H., Triandis, H. C., and Yee, C. (1991). Cultural differences in reward allocation: is collectivism the explanation? *British Journal of Social Psychology, 30*, 145–157.

Hull, T. H. (1988). Charity begins at home? Attitudes toward foreign aid. In J. Kelley and C. Bean (eds), *Australian Attitudes* (pp. 36–44). Sydney: Allen & Unwin.

Huseman, R. C., Hatfield, J. D., and Miles, E. W. (1987). A new perspective on Equity theory: the Equity Sensitivity construct. *Academy of Management Review, 12*, 222–234.

Hyma, B. and Ramesh, A. (1994). Traditional medicine: its extent and potential for incorporation into modern national health systems. In D. R. Phillips and Y. Verhasselt (eds), *Health and Development* (pp. 65–82). London: Routledge.

ICFTU (International Confederation of Free Trade Unions), Zambia Congress of Trade Unions and Friedrich Ebert Stiftung (1992). 'The Social Dimensions of Adjustment in Zambia'. Background document, Conference in Lusaka, 18–20 November.

Ichheiser, G. (1943). Misunderstandings of personality in everyday life and the psychologist's frame of reference. *Character and Personality, 12*, 145–160.

IMF Report (1992). IMF and World Bank staff exchange views with labour union officials at seminar. IMF Survey, Washington, DC, 30 November, p. 359.

Ingram, J. (former UN Under-Secretary and Director of the World Food Programme) (1994). Aid backlash feared. *Development Bulletin, 30*, 61.

INRA. (1992). *The Way Europeans Perceive the Third World in 1991*. Brussels: European Community Commission Report.

Isen, A. M., and Noonberg, A. (1979). The effect of photographs of the handicapped on donation to charity: when a thousand words may be too much. *Journal of Applied Social Psychology, 9*, 4260431.

Isenberg, D. J. (1986). Group polarization: a critical review and meta-analysis. *Journal of Personality and Social Psychology, 6*, 1141–1151.

Jahode, G. (1970). Supernatural beliefs and changing cognitive structures among Ghanaian University students. *Journal of Cross-cultural Psychology, 1(2)*, 115–130.

Janis, I. (1982). *Groupthink: Psychological Studies of Policy Decisions and Fiascoes.* Boston, MA: Houghton Mifflin.

Janis, I. L., and Feshbach, S. (1953). Effects of fear-arousing communications. *Journal of Abnormal and Social Psychology, 48*, 78–92.

Jeshmaridian, S., and Takooshian, H. (1994). Country profile: Armenia. *Psychology International, 8*, 9.

Jin, P. (1993). Work motivation and productivity in voluntarily formed work teams. *Organizational Behavior and Human Decision Processes, 54*, 133–155.

Johnson, D. B. (1982). The free-rider principle, the charity market, and the economics of blood. *British Journal of Social Psychology, 21*, 93–106.

Johnson, D. W., and Johnson, F. P. (1990). *Joining Together: Group Theory and Group Skills.* Englewood Cliffs, NJ: Prentice-Hall.

Jones, B. (1990). An essential ingredient in short supply. *Human Resources Monthly, July*, 6.

Jones, D., Reese, A., and Walker, C. (1994). Finding a role for psychology in development. *Why Psychology? Australian Psychological Society Annual Conference Abstracts, 29*, 57.

Jones, E. E., and Harris, V. A. (1967). The attribution of attitudes. *Journal of Experimental Social Psychology, 3*, 1–24.

Jones, E. E., and Nisbett, R. E. (1972). The actor and the observer: divergent perceptions of the causes of behavior. In E. E. Jones, D. E. Kanouse, H. H. Kelley, R. E. Nisbett, S. Valins, and B. Weiner (eds), *Attribution: Perceiving the Causes of Behavior* (pp. 79–94). Morristown, NJ: General Learning Press.

Jones, M. (1988). Managerial thinking: an African perspective. *Journal of Management Studies, 25*, 481–505.

Jones, M. (1991). Management development: an African focus. In M. Mendenhall, and G. Oddou (eds), *Research and Cases in Human Resource Management* (pp. 231–244). Boston, MA: PWS Kent Publishing Company.

Jones, P., and Barr, A. (1996). Learning by doing in sub–Saharan Africa: evidence from Ghana. *Journal of International Development, 8*, 445–466.

Jones, R., and Popper, R. (1972). Characteristics of Peace Corps host countries and the behavior of volunteers. *Journal of Cross-Cultural Psychology, 3*, 233–245.

Jones, S., Carr, S. C., and Casimir, G. (1995). Work motivation down under: labouring under an illusion? *Journal of Psychology and the Behavioral Sciences, 9*, 85–103.

Jordan, P. C. (1986). Effects of an extrinsic reward on intrinsic motivation: a field experiment. *Academy of Management Journal, 29*, 405–412.

Joyce, R. E., and Hunt, C. L. (1982). Philippine nurses and the brain drain. *Social Science and Medicine, 16*, 1223–1233.

Juralewicz, R. S. (1974). An experiment in participation in a Latin American factory. *Human Relations, 27*, 627–637.

Kâğitçibaşi, Ç. (1995). Is psychology relevant to global human development issues? Experience from Turkey. *American Psychologist, 50*, 293–300.

Kanyangale, M., and MacLachlan, M. (1995). Critical incidents for refugee counsellors: an investigation of indigenous human resources. *Counselling Psychology Quarterly, 8*, 89–101.

Kao, H. S. R., and Ng, Sek-Hong (1996). Work motivation and culture. In D. Munro, J. F. Schumaker, and S. C. Carr (eds), *Motivation and Culture* (pp. 119–132). New York: Routledge.

Kaplinsky, R. (1995). Technique and system: the spread of Japanese management techniques to developing countries. *World Development, 23*, 57–71.

Karpin, D. (1995). *Enterprising Nation: Renewing Australia's Managers to Meet the Challenges of the Asia–Pacific Century*. Canberra: Australian Government Publishing Services.

Kasente, D. H. (1996). Gender studies and gender training in Africa. *Development Bulletin, 37*, 30–33.

Kashima, Y. (1997). Culture, narrative, and human motivation. In D. Munro, J. F. Schumaker, and S. C. Carr (eds), *Motivation and Culture* (pp. 16–30). New York: Routledge.

Kashima, Y., and Callan, V. J. (1994). The Japanese work group. In H. C. Triandis, M. D. Dunnette, and L. M. Hough (eds), *Handbook of Industrial and Organizational Psychology, Vol. 4* (pp. 609–646). Palo Alto, CA: Consulting Psychologists Press.

Kaul, M. (1989). Strategic issues in development management: learning from successful experiences. *International Journal of Public Sector Management, 1*, 12–25.

Kaur, R., and Ward, C. (1992). Cross-cultural construct validity study of 'fear of success': a Singaporean case study. In S. Iwawaki, Y. Kashima, and K. Leung (eds), *Innovations in Cross-cultural Psychology* (pp. 214–222). Lisse: Swets & Zeitlinger.

Kealey, D. J. (1989). A study of cross-cultural effectiveness: theoretical issues, practical applications. *International Journal of Intercultural Relations, 13*, 387–428.

Keats, D. M. (1969). *Back in Asia: A Follow-Up Study of Australian-Trained Asian Students.* Canberra: Australian National University Press.

Keats, D. M. (1993a). *Skilled Interviewing.* Melbourne: ACER (Australian Council for Educational Research).

Keats, D. M., and Fu-Xi, F. (1996). The development of concepts of fairness in rewards in Chinese and Australian children. In H. Grad, A. Blanco, and J. Georgas (eds), *Key Issues in Cross-cultural Psychology* (pp. 276–287). Lisse, NL: Swets & Zeitlinger.

Kelley, H. H. (1972). Causal schemata and the attribution process. In E. E. Jones, D. E. Kanouse, H. H. Kelley, R. E. Nisbett, S. Valins, and B. Weiner (eds), *Attribution: Perceiving the Causes of Behavior* (pp. 151–174). Morristown, NJ: General Learning Press.

Kelley, J. C. (1989). Australian attitudes to overseas aid: report from the National Science Survey. *International Development Issues, 8*, 1–129.

Kelman, H.C (1974). Attitudes are alive and well and gainfully employed in the sphere of action. *American Psychologist, 29*, 310–335.

Kettley, P. (1997). *Personal Feedback: Cases in Point.* Brighton: Institute for Employment Studies.

Kiggundu, M. N. (1986). Limitations to the application of sociotechnical systems in developing countries. *Journal of Applied Behavioural Science, 22*, 341–353.

Kiggundu, M. N. (1990). Limitations to the application of socio-technical systems in developing countries. In A. M. Jaeger, and R. N. Kanungo (eds), *Management Development in Developing Countries* (pp. 146–162). Chippenham: Routledge.

Kiggundu, M. N. (1991). The challenge of management development in Sub-Saharan Africa. *Journal of Management Development, 10*, 32–47.

Kim, U. (1994). *Rediscovering the Human Mind: The Indigenous Psychologies Approach with Specific Focus on East Asian Cultures.* Kuala Lumpur: Universiti Kebangsaan Malaysia.

Kinder, R. (1994). *Indonesia Land Admininstration Project: Selecting Applicants for Scholarships.* Jakarta: IDP (International Development Program) Education Australia Ltd.

King, W. C., Miles, E. W., and Day, D. D. (1993). A test and refinement of the equity sensitivity construct. *Journal of Organizational Behavior, 14*, 301–317.

Kipnis, D. (1972). Does power corrupt? *Journal of Personality and Social Psychology, 24*, 33–41.

Kipnis, D. (1977). *The Powerholders.* Chicago: University of Chicago Press.

Kishindo, P. (1996). Sexual behaviour in the face of risk: the case of bar girls in Malaẅi's major cities. *Health Transition Review, 5* (supplement), 153–160.

Kleiner, K. (1996). West bows to psychology's cultural revolution. *New Scientist, Feb.*, 12.

Klitgaard, R. (1995). Including culture in evaluation research. *New Directions for Evaluation*, *67*, 135–146.

Kloos, H. (1990). Health aspects of resettlement in Ethiopia. *Social Science and Medicine*, *30*, 643–656.

Kniveton, B. (1989). *The Psychology of Bargaining*. Aldershot: Gower Publishing Company.

Kohnert, D. (1996). Magic and witchcraft: implications for democratization and poverty-alleviating aid in Africa. *World Development*, *24*, 1347–1355.

Kolb, D. A. (1984). *Experiential Learning*. Englewood Cliffs, NJ: Prentice-Hall.

Konkoly, T. H., and Perloff, R. M. (1990). Applying the theory of reasoned action to charitable intent. *Psychological Reports*, *67*, 91–94.

Korten, F. F. (1974). The influence of culture and sex on the perception of persons. *International Journal of Psychology*, *9*, 31–44.

Kramer, R. M., and Messick, D. M. (eds) (1995). *Negotiation as a Social Process*. Thousand Oaks, CA: Sage.

Kramer, R. M., Shah, P. P., and Woerner, S. L. (1995). Why ultimatums fail: social identity and moralistic aggression in coercive bargaining. In R. M. Kramer and D. M. Messick (eds), *Negotiation as a Social Process* (pp. 285–308). Thousand Oaks, CA: Sage.

Krewer, B., and Jahoda, G. (1993). Psychologie et culture: Vers une solution du 'Babel'? *International Journal of Psychology*, *28*, 367–375.

Krishnan, L., and Carment, D. W. (1981). The effect of amount of help, resource control, donor's sex, and recipient's sex, on acceptance of aid. *Psychologia: An International Journal of Psychology in the Orient*, *24*, 14–20.

Kumar, K. (1995). Measuring the performance of agricultural and rural development programs. *New Directions for Evaluation*, *67*, 81–91.

Kumar, R. (1995). Including culture in evaluation research. *New Directions for Evaluation*, *67*, 135–146.

Laford, A., and Seaman, J. (1991/2). Sustainability in the health sector. *The Health Exchange*, p. 3.

Lamarche, L., and Tougas, F. (1979). Perception des raisons de la pauvreté par des Montréalais Canadiens-Français. *Revue Canadienne des sciences du comportement*, *11*, 72–78.

Lamm, H., and Myers, D. G. (1978). Group induced polarization of attitudes and behaviour. *Advances in Experimental Social Psychology*, *11*, 145–187.

Landy, F. J. (1989). *Psychology of Work Behavior*. Pacific Grove, CA: Brooks/Cole.

Lane, H. W. and Burgoyne, D. G. (1988). Management development in Africa: the Canada–Kenya Executive Management Programme experience. *Journal of Management Development*, *7*, 40–55.

La Sierra, U. (1992). Beliefs, value orientation, and culture in attribution processes and helping behavior. *Journal of Cross-Cultural Psychology*, *23*, 179–195.

Lawler, E. E., Kopler, C. A., Young, T. F., and Fadem, J. A. (1968). Inequity reduction over time in an induced overpayment situation. *Organizational Behavior and Human Performance*, *3*, 253–268.

Lawrence, J. E. S. (1989). Engaging recipients in development evaluation: the 'stakeholder' approach. *Evaluation Review*, *13*, 243–256.

Lawuyi, O. B. (1992). Vehicle slogans as personal and social thought: a perspective on self development in Nigeria. *New Directions for Educational Reform*, *1*, 91–98.

Leavitt, H. J. (1951). Some effects of certain communication patterns on group performances. *Journal of Abnormal and Social Psychology*, 46, 38–50.

Leeds, R. (1963). Alturism and the norm of giving. *Merrill-Palmer Quarterly of Behavior and Development*, 9, 229–240.

Lerner, M. J. (1970). The desire for justice and reaction to victims. In J. Macauley and L. Berkowitz (eds), *Altruism and Helping Behavior*. New York: Academic Press.

Lerner, M. J. (1980). *The Belief in a Just World: A Fundamental Delusion*. New York: Plenum Press.

Lerner, M. J., and Miller, D. T. (1978). Just world research and the attribution process: looking back and ahead. *Psychological Bulletin*, 85, 1030–1051.

Lerner, M. J., and Simmons, C. H. (1966). Observer's reaction to the 'innocent victim': compassion or rejection? *Journal of Personality and Social Psychology*, 4, 203–210.

Leung, K., and Bond, M. H. (1984). The impact of cultural collectivism on reward allocation. *Journal of Personality and Social Psychology*, 47, 793–804.

Leung, K., and Park, H. J. (1986). Effects of interactional goal on choice of allocation rules: a cross-national study. *Organizational Behavior and Human Decision Processes*, 37, 111–120.

Leung, K., and Wu, P. G. (1990). Dispute processing: a cross-cultural analysis. In R. W. Brislin (ed.), *Applied Cross-Cultural Psychology* (pp. 209–231). Newbury Park, CA: Sage.

Level, D. A., Ormsby, J. E., Watts, L. R., and Tinsley, D. B. (1990). Management by objectives: implications for managerial communication. *Journal of Managerial Issues*, 2, 325–336.

Leventhal, H. (1973). Fear communication in the acceptance of preventive health practices. *New York Academy of Medicine*, 41, 1144–1168.

Levine, D. I. (1993). What do wages buy? *Administrative Science Quarterly*, 38, 462–483.

Lewin, K. (1952). Group decisions and social change. In T. Newcomb and T. Hartley (eds), *Readings in Social Psychology*. New York: Holt, Rhinehart and Wilson.

Lewin, K., Lippitt, R., and White, R. K. (1939). Patterns of aggressive behavior in experimentally created 'social climates'. *Journal of Social Psychology*, 10, 271–299.

Lewis, J. (1991). Revealuating the effect of N.Ach on economic growth. *World Development*, 19, 1269–1274.

Linder, D. E., Jones, E. E., and Cooper, J. (1967). Decision freedom as a determinant of the role of incentive magnitude in attitude change. *Journal of Personality and Social Psychology* 6, 245–254.

Linton, S. J., and Warg, L. E. (1993). Attributions (beliefs) and job satisfaction associated with backpain in an industrial setting. *Perceptual and Motor Skills*, 76, 51–62.

Liomba, N. G. (1994). *Statistics for HIV/AIDS in Malaŵi*. Lilongwe, Malaŵi: Ministry of Health.

Littlewood, R. (1985). Jungle madness: some observations on expatriate psychopathology. *International Journal of Social Psychiatry*, 31, 194–197.

Locke, E. A. (1968). Toward a theory of task motivation and incentives. *Organizational Behavior and Human Performance*, 3, 157–189.

Lopez, S. A. (1987). Sources of chronic rural poverty: behavioral perspectives. *Revista interamericana de psicología*, 21, 72–89.

Lynskey, M. T., Ward, C., and Fletcher, G. J. (1991). Stereotypes and intergroup attributions in New Zealand. *Psychology and Developing Societies, 3*, 113–127.

Machlup, F. (1962). *The Production and Distribution of Knowledge in the United States.* Princeton, NJ: Princeton University Press.

Machungwa, P., and Schmitt, N. (1983). Work motivation in a developing country. *Journal of Applied Psychology, 68*, 31–42.

MacLachlan, M. (1992). Debriefing in brief. *Training and Development, 24*, 23–25.

MacLachlan, M. (1993a). Splitting the difference: how do refugee workers survive? *Changes: International Journal of Psychology and Psychotherapy, 11*, 155–157.

MacLachlan, M. (1993b). Sustaining human resource development in Africa: the influence of expatriates. *Management Education and Development, 24*, 153–157.

MacLachlan, M. (1993c). Sustaining health service developments in the 'Third World'. *Journal of the Royal Society of Health, 113*(3), 136–137.

MacLachlan, M. (1994). Tropical health promotion: On learning from the community. International Congress of Applied Psychology, Madrid, July.

MacLachlan, M. (1996a). From sustainable change to incremental improvement: The psychology of community rehabilitation. In S. C. Carr and J. F. Schumaker (eds), *Psychology and the Developing World* (pp. 26–37). Westport, CT: Praeger.

MacLachlan, M. (1996b). Creating a psychology for development. *Changes: International Journal of Psychology and Psychotherapy, 14*(2), 142–147.

MacLachlan, M. (1996c). Identifying problems in community health promotion: an illustration of the Nominal Group Technique in AIDS education. *Journal of the Royal Society of Health, 116*(3), 143–148.

MacLachlan, M. (1997). *Culture and Health.* Chichester: John Wiley.

MacLachlan, M., Banda, D. M., and McAuliffe, E. (1995c). Epidemic psychological disturbance in a Malaŵian secondary school: a case study in social change? *Psychology and Developing Societies, 7*, 79–90.

MacLachlan, M., and Carr, S. C. (1993a). Demotivating the doctors: the double demotivation hypothesis in Third World health services. *Journal of Management in Medicine, 7*, 6–10.

MacLachlan, M., and Carr, S. C. (1993b). Marketing psychology in a developing country: an innovative application in Malaŵi. *Psychology Teaching Review, 2*, 22–29.

MacLachlan, M., and Carr, S. C. (1994). Pathways to a psychology for development: reconstituting, restating, refuting, and realizing. *Psychology and Developing Societies, 6*, 21–28.

MacLachlan, M., and Carr, S. C. (1994a). From dissonance to tolerance: towards managing health in tropical cultures. *Psychology and Developing Societies, 6*, 119–129.

MacLachlan, M., and Carr, S. C. (1994b). Managing the AIDS Crisis in Africa: In support of pluralism. *Journal of Management in Medicine, 8*, 45–52.

MacLachlan, M., and Carr, S. C. (1997). Psychology in Malaŵi: towards a constructive debate. *The Psychologist, 10*, 77–79.

MacLachlan, M., Carr, S. C., Fardell, S., Maffesoni, G., and Cunningham, J. (1997). Transactional analysis of communication styles in HIV/AIDS advertisements. *Journal of Health Psychology, 2*, 67–74.

MacLachlan, M., Chimombo, M., and Mpemba, N. (1996). AIDS education for youth through active learning: a school–based approach from Malaŵi. *International Journal of Educational Development, 16*, 1–10.

MacLachlan, M., and McAuliffe, E. (1993). Critical incidents for psychology students in a refugee camp: implications for counselling. *Counselling Psychology Quarterly*, 6, 3–11.

MacLachlan, M., Nyirenda, T., and Nyando, C. (1995b). Attributions for admission to Zomba Mental Hospital: implications for the development of mental health services in Malawi. *International Journal of Social Psychiatry*, 41, 79–87.

MacLachlan, M., Page, R., Robinson, G. L., Nyirenda, T. and Ali, S. (1998). Patient's perceptions of chamba (marijuana) use in Malawi. *Substance Use and Misuse*, 33 (6), 1–7.

Madill, H. M. P., Brintnell, S. G., and Tjandrakasuma, H. (1995). Forging educational links between Canada and Indonesia: implementing programs in allied health. In S. Aroni and T. Adams (eds), *Proceedings of the Fourth International Symposium on the Role of Universities in Developing Areas Vol. 1* (pp. 1–18). Melbourne: INRUDA/UNESCO/IDTC/RMIT.

Makin, P., Cooper, C., and Cox, C. (1989). *Managing People at Work*. London: Routledge.

Mamman, A. (1994). *Expatriate's Identity in a Global Village: Implications and Propositions*. Third World Business Congress Proceedings (pp. 386–397). Penang, Malaysia: International Management Development Association.

Mamman, A. (1995). Expatriate adjustment: Dealing with hosts' attitudes in a foreign assignment. *Journal of Transnational Management Development*, 1, 49–70.

Manning, M. R., and Avolio, B. J. (1985). The impact of pay disclosure in a university environment. *Research in Higher Education*, 23, 135–149.

Mansell, D. (1995). Technology transfer. *Asian Studies Review*, 18, 17–27.

Mansell, D. S. (1978). Engineering standards in developing countries. *Engineering Issues – Journal of Professional Activities*, ACSE, 104 (paper 14103), 271–291.

Mansell, D. S. (1987). *Risk – A Vehicle for Change, but a Barrier to Change*. Proceedings of the UNESCO Regional Seminars for South East Asia and the Pacific on Barriers to Change and Technical Implications for the Provision of Shelter for the Homeless by the Year 2000, November. Melbourne: University of Melbourne.

Mansell, D. S. (1995). Technology Transfer. *Asian Studies Review*, 18(3), 17–27.

Marai, L. (1996). The development of psychology in Papua New Guinea: A brief review. *South Pacific Journal of Psychology*, 9, 1–6.

Marin, G. (1985). The preference for equity when judging the attractiveness and fairness of an allocator: the role of familiarity and culture. *Journal of Social Psychology*, 125, 543–549.

Marjoribanks, K., and Jordan, D. F. (1986). Stereotyping among Aboriginal and Anglo-Australians: the uniformity, intensity, direction, and quality of auto- and heterostereotypes. *Journal of Cross-Cultural Psychology*, 17, 17–28.

Markus, H., and Kitayana, S. (1991). Culture and the self: implications for cognition, emotion, and motivation. *Psychological Review*, 98, 224–253.

Marrow, A. J. (1964). Risks and uncertainties in action research. *Journal of Social Issues*, 20, 5–20.

Marsella, A. J., and Choi, S. C. (1993). Psychosocial aspects of modernization and economic development in East Asian nations. *Psychologia: An International Journal of Psychology in the Orient*, 36, 201–213.

Marsella, A. J., Levi, L., and Ekblad, S. (1996). The importance of including Quality of Life indices in international and social economic development activities. *Applied and Preventive Psychology*, *6*, 55–67.

Martinko, M. (ed). (1994). *Attribution Theory: An Organizational Perspective*. Tampa, FL: St Lucie Press.

Marx, K., and Engels, F. (1948). *The Communist Manifesto*. London: Lawrence & Wishart (centenary edition).

Maslach, C. (1982). *Burnout: The Cost of Caring*. London: Prentice-Hall.

Maslow, A. H. (1954). *Motivation and Personality*. New York: Harper & Row.

Mathur, A. (1983). Generosity and reciprocity in a rural Indian setting. *Journal of Social Psychology*, *121*, 147–148.

Matsuda, N. (1985). Strong, quasi- and weak conformity among Japanese in the modified Asch procedure. *Journal of Cross-Cultural Psychology*, *16*, 83–97.

Mayo, E. (1949). *The Social Problems of an Industrial Civilization*. New York: Routledge.

Mazrui, A. A. (1980). *The African Condition*. London: Heinemann.

McArthur, L. Z., and Post, D. L. (1977). Figural emphasis and person perception. *Journal of Experimental Social Psychology*, *13*, 520–535.

McAuliffe, E. (1994a). *An Evaluation of a UNICEF Social Mobilisation Programme in the Southern Region of Malawi*. Centre for Social Research, University of Malawi.

McAuliffe, E. (1994b). *AIDS: The Barriers to Behaviour Change*. Lilongwe, Malawi: UNICEF.

McAuliffe, E. (1996). AIDS: Barriers to behavior change in Malawi. In H. Grad, A. Blanco, and J. Georgas (eds), *Key Issues in Cross-Cultural Psychology* (pp. 371–386). Amsterdam: Swets & Zeitlinger.

McClelland, D. C. (1961). *The Achieving Society*. Princeton, NJ: D. Van Nostrand and Company.

McClelland, D. C. (1987). *Human motivation*. Cambridge: Cambridge University Press.

McClelland, D., and Winter, D. G. (1971). *Motivating Economic Achievement*. New York: Free Press.

McGregor, D. (1960). *The Human Side of Enterprise*. New York: McGraw-Hill.

McGuire, W. J. (1985). Attitudes and attitude change. In G. Lindzey and E. Aronson (eds), *Handbook of Social Psychology*, *Vol. 2* (pp. 233–346). New York: Random House.

McKee, N. (1992). *Social Mobilization and Social Marketing in Developing Communities: Lessons for Communicators*. Penang, Malaysia: Southbound.

McLoughlin, D., and Carr, S. C. (1994). *The Buick Bar & Grill*. Melbourne, Australia: University of Melbourne Case Study Services.

McLoughlin, D., and Carr, S. C. (1997). Equity sensitivity and double demotivation. *Journal of Social Psychology*, *137*, 668–670.

McManus, I. C., Richards, P., Winder, B. C., Sproston, K. A., and Styles, V. (1995). Medical school applicants from ethnic minority groups: identifying if and when they are disadvantaged. *British Medical Journal*, *310*, 496–500.

Meaning of Work (MOW) International Research Team. (1987). *The Meaning of Work: An International View*. London: Academic Press.

Mehryar, A. H. (1984). The role of psychology in national development: wishful thinking and reality. *International Journal of Psychology*, *19*, 159–167.

Meisner, C. (1996). Family training in Bangladesh: an innovative approach. *Development Bulletin, 37,* 57–58.

Melikian, L. H. (1984). The transfer of psychological knowledge to the Third World countries and its impact on development: the case of five arab Gulf oil-producing states. *International Journal of Psychology, 19,* 65–67.

Memmi, A. (1966). *Portrait du colonisé.* Utrecht: Jean-Jacques Pauvert.

Mendenhall, M. E. and Wiley, C. (1994). Strangers in a strange land. *American Behavioral Scientist, 37,* 605–620.

Merton, R. K. (ed.). (1968). *Social Theory and Social Structure.* New York: Free Press.

Meyer, J. P., and Mulherin, A. (1980). From attribution to helping: an analysis of the mediating effects of affect and expectancy. *Journal of Personality and Social Psychology, 39,* 201–210.

Miller, J. G. (1984). Culture and the development of everyday social explanation. *Journal of Personality and Social Psychology, 46,* 961–978.

Miller, J. G., Bersoff, D. M., and Harwood, R. L. (1990). Perceptions of social responsibilities in India and in the United States: moral imperatives or personal decisions? *Journal of Personality and Social Psychology, 58,* 33–47.

Mills, G. G. (1994). Community development and fish farming in Malawi. *Community Development Journal, 29,* 215–221.

Misumi, J., and Peterson, M. F. (1985). The Performance-Maintenance (PM) theory of leadership: review of Japanese research program. *Administrative Science Quarterly, 30,* 198–223.

Mitchell, T. R., and Kalb, L. S. (1982). Effects of job experience on supervisor's attributions for a subordinate's poor performance. *Journal of Applied Psychology, 67,* 181–188.

Moghaddam, F. M. (1989). Specialization and despecialization in psychology: divergent processes in the three worlds. *International Journal of Psychology, 24,* 103–116.

Moghaddam, F. M. (1990). Modulative and generative orientations in psychology: implications for psychology in the three worlds. *Journal of Social Issues, 46,* 21–41.

Moghaddam, F. M. (1993). Traditional and modern psychologies in competing cultural systems: lessons from Iran 1978–1981. In U. Kim and J. W. Berry (eds), *Indigenous Psychologies: Research and Experience in Cultural Context.* Newbury Park, CA: Sage.

Moghaddam, F. M. (1996). Training for developing world psychologists: can it be better than the psychology? In S. C. Carr and J. F. Schumaker (eds), *Psychology and the Developing World* (pp. 49–59). Westport, CT: Praeger.

Moghaddam, F. M., and Taylor, D. M. (1986). What constitutes an 'appropriate psychology' for the developing world? *International Journal of Psychology, 21,* 253–267.

Moghaddam, F. M., and Taylor, D. M. (1987). Towards appropriate training for developing world psychologists. In Ç. Kâğitçibaşi (ed.), *Growth and Progress in Cross-Cultural Psychology* (pp. 69–75). Lisse: Swets & Zeitlinger.

Monan, J., and Porter, D. (1996). Agricultural extension services in Vietnam: some legacies and prospects. *Development Bulletin, 37,* 49–51.

Monk, G., Winslade, J., Crocket, K., and Epston, E. (1996). *Narrative Therapy in Practice.* New York: Jossey-Bass.

231

Monson, T. C., and Snyder, M. (1977). Actors, observers, and the attribution process: towards a reconceptualization. *Journal of Experimental Social Psychology*, 13, 89–111.

Montero, M. (1990). Ideology and psycho-social research in Third World contexts. *Journal of Social Issues*, 46, 43–56.

Moore, J., Becker, C., Harrison, J. S., and Eng, T. R. (1995). HIV risk behavior among Peace Corps volunteers. *AIDS*, 9, 795–799.

Morris, M. W., and Peng, K. (1994). Culture and cause: American and Chinese attributions for social and physical events. *Journal of Personality and Social Psychology*, 67, 949–971.

Morris, M. W., Sim, D. L. H., and Girotto, V. (1995). Time of decision, ethical obligation, and causal illusion. In R. M. Kramer and D. M. Messick (eds), *Negotiation as a Social Process* (pp. 209–239). Thousand Oaks, CA: Sage.

Morse, N. C., and Weiss, R. S. (1955). The function and meaning of work and the job. *American Sociological Review*, 20, 191–198.

Moscovici, S. (1976). *Social Influence and Social Change*. London: Academic Press.

Moscovici, S. (1980). Toward a theory of conversion behavior. *Advances in Experimental Social Psychology*, 13, 209–239.

Moscovici, S., and Zavalloni, M. (1969). The group as a polarizer of attitudes. *Journal of Personality and Social Psychology*, 12, 125–135.

Muchinsky, P. M. (1993). *Psychology Applied to Work*. Pacific Grove, CA: Brooks/Cole.

Muchinsky, P. M. (1997). *Psychology Applied to Work*. Pacific Grove, CA: Brooks/Cole.

Munandar, A. S. (1990). Indonesian managers, today and tomorrow. *International Journal of Psychology*, 25, 855–869.

Munene, J. C. (1995). 'Not-on-seat': an investigation of some correlates of organisational citizenship behaviour in Nigeria. *Applied Psychology: An International Review*, 44, 111–122.

Munro, D. (1996). Work motivation and developing nations: a systems approach. In S. C. Carr, and J. F. Schumaker (eds), *Psychology and the Developing World* (pp. 130–139). Westport, CT: Praeger.

Murnighan, J. K., and Pillutla, M. M. (1995). Fairness versus self-interest. In R. M. Kramer and D. M. Messick (eds), *Negotiation as a Social Process* (pp. 240–267). Thousand Oaks, CA: Sage.

Murray, M. (1997). A narrative approach to health psychology: background and potential. *Journal of Health Psychology*, 2, 9–20.

Mwaniki, N. (1996). Rural development in Kenya: an overview of the gap between policy and practice. *Development Bulletin*, 37, 15–19.

Myers, D. G. (1975). Discussion-induced attitude polarization. *Human Relations*, 28, 699–714.

Myers, D. G., and Bishop, G. D. (1970). Discussion effects on racial attitudes. *Science*, 169, 778–779.

Nadler, A., Fisher, J. D., and Streufert, S. (1974). The donor's dilemma: recipient's reaction to aid from friend or foe. *Journal of Applied Social Psychology*, 4, 275–285.

Nadler, A., and Mayseless, O. (1983). Recipient self-esteem and reactions to help. In J. D. Fisher, A. Nadler, and B. M. DePaulo (eds), *New Directions in Helping, Vol 1* (pp. 167–188). New York: Academic Press.

Nash, J. F. (1950). The bargaining problem. *Econometrica*, 18, 155–162.

Nasir, R., and Ismail, N. (1994). Attitudes toward women managers and its relationship with fear of success among female students. Kuala Lumpur: Third Afro-Asian Psychological Congress, August.

Nelson, J. (1991). Organized labour, politics and labour market flexibility in developing countries. *The World Bank Research Observer*, Washington, DC, 6(1) (Jan.), 53.

Newstead, S. E. (1992). The use of examinations in the assessment of psychology students. *Psychology Teaching Review*, 1, 22–33.

Newstead, S. E. (1996). The psychology of student assessment. *The Psychologist*, 9, 543–547.

Nichter, M. (1955). Vaccinations in the Third World: a consideration of community demand. *Social Science and Medicine*, 41, 617–632.

Nisbett, R. E., Caputo, C., Legant, P., and Maracek, J. (1973). Behavior as seen by the actor and as seen by the observer. *Journal of Personality and Social Psychology*, 27, 154–164.

Nisbett, R. E., and Ross, L. (1980). *Human Inference: Strategies and Shortcomings of Social Judgments*. Englewood Cliffs, NJ: Prentice-Hall.

Nixon, M. (1994). Practices and needs in psychological training: a survey of 28 countries. *Australian Psychologist*, 29, 166–173.

Nyirenda, D., and Jere, D. R. (1991). *An Evaluation Report for AIDS Education Materials*. Domasi: Malaŵi Institute of Education.

Nykodym, N., Longenecker, C. O., and Ruud, W. N. (1991). Improving quality of life with transactional analysis as an intervention change strategy. *Applied Psychology: An International Review*, 40, 395–404.

O'Dwyer, T., and Woodhouse, T. (1996). The motivations of Irish Third World development workers. *Irish Journal of Psychology*, 17, 23–34.

Oakland, T. (1995). 44-country survey shows international test use patterns. *Psychology International*, 6, 7.

Ong, C. N. (1991). Ergonomics, technology transfer, and developing countries. *Ergonomics*, 34, 799–814.

Orpen, C. (1990). Measuring support for organizational innovation: A validity study. *Psychological Reports*, 67, 417–418.

Orpen, C. (1993). The Multifactorial Achievement Scale as a predictor of salary growth and motivation among middle managers. *Psychological Studies*, 38, 79–81.

Overseas Development Administration (ODA). (1995). Personal correspondence, 10 May.

Owusu-Bempah, J., and Howitt, D. (1995). How Eurocentric psychology damages Africa. *The Psychologist*, 10, 462–465.

Paige, R. M. (1990). International students: cross-cultural psychological perspectives. In R. W. Brislin (ed.), *Applied Cross-Cultural Psychology* (pp. 161–185). Newbury Park, CA: Sage.

Pandey, J., Sinha, Y., Prakash, A., and Tripathi, R. C. (1982). Right–left political ideologies and attribution of the causes of poverty. *European Journal of Social Psychology*, 12, 327–331.

Pangani, D., Carr, S. C., MacLachlan, M., and Ager, A. (1993). Medical versus traditional attributions for psychiatric symptoms in the tropics: which is associated with greater tolerance? *Medical Science Research*, 21(4), 171.

Paranjpe, A. C. (1994). Merits and limits of East/West integration of concepts and remedies of suffering. 23rd International Congress of Applied Psychology, Madrid, July.

Park, B., and Rothbart, M. (1982). Perception of outgroup homogeneity and levels of social categorization: Memory for the subordinate attributes of ingroup and outgroup members. *Journal of Personality and Social Psychology*, 42, 1051–1068.

Parker, B., and McEvoy, G. M. (1993). Initial examination of a model of intercultural adjustment. *International Journal of Intercultural Relations*, 17, 355–379.

Patchen, M. (1959). *Study of Work and Life Satisfaction: Absences and Attitudes towards Work Experience*. Ann Arbor, MI: Ann Arbor Institute for Social Research.

Paul, E. (1996). Towards a new Cold War? *Development Bulletin*, 37, 37–38.

Paul, S. (1995). Evaluating public services: A case study fro Bangalore, India. *New Directions for Evaluation*, 67, 154–165.

Payer, (1975). *The Debt Trap: The IMF and the Third World*. Harmondsworth: Penguin.

Payne, M., and Furnham, A. (1985). Explaining the causes of poverty in the West Indies: a cross-cultural comparison. *Journal of Economic Psychology*, 6, 215–229.

Pearce, F. (1996). Squatters take control. *New Scientist, 1 June*, 38–42.

Peltzer, K. (1989). Some contributions to the psychology of traditional healing in Malawi etc.

Penna-Firme, T., Grinder, R. E., and Linhares-Barreto. (1991). Adolescent female prostitutes on the streets of Brazil: an exploratory investigation of ontological issues. *Journal of Adolescent Research*, 6, 493–504.

Perry, L. S. (1993). Effects of inequity on job satisfaction and self-evaluation in a national sample of African-American workers. *Journal of Social Psychology*, 133, 565–573.

Pettigrew, T. F. (1979). The ultimate attribution error: extending Allport's cognitive analyses of prejudice. *Personality and Social Psychology Bulletin*, 5, 461–476.

Pfost, K. S., and Fiore, M. (1990). Pursuit of nontraditional occupations: fear of success or fear of not being chosen. *Sex Roles*, 23, 15–24.

Picciotto, R., and Rist, R. C. (eds). (1995). Evaluation and development. *New Directions for Evaluation*, 67 (whole issue), 1–175.

Porteous, M. (1997). *Occupational Psychology*. London: Prentice-Hall.

Porter, D., Allen, B., and Thompson, G. (1991). *Development in Practice: Paved with Good Intentions*. London: Routledge.

Posthuma, A. C. (1995). Japanese techniques in Africa? Human resources and industrial restructuring in Zimbabwe. *World Development*, 23, 103–116.

Power, V. (1994). *Gender Bias and the 'Glass Ceiling'*. 29th Annual Conference of the Australian Psychological Society, Wollongong, October.

Powis, D. A., Neame, R. L. B., Bristow, T., and Murphy, L. B. (1988). The objective structured interview for medical student selection. *British Medical Journal*, 296, 765–768.

Pruitt, D. G. (1981). *Negotiation Behavior*. New York: Academic Press.

Pruitt, D. G. (1995). Networks and collective scripts: paying attention to structure in bargaining theory. In R. M. Kramer and D M. Messick (eds), *Negotiation as a Social Process* (pp. 37–47). Thousand Oaks, CA: Sage.

Pryor, R. G. L. (1993). Returning from the wilderness. *Australian Journal of Career development, September*, 13–17.

Psacharopoulos, G. (1995). Using evaluation indicators to track the performance of education programs. *New Directions for Evaluation*, 67, 93–104.

Punnett, B. J. (1986). Goal setting: an extension of the research. *Journal of Applied Psychology*, 71, 171–172.

Radley, A., and Kennedy, M. (1992). Reflections upon charitable giving: a comparison of individuals from business, 'manual,' and professional backgrounds. *Journal of Community and Applied Social Psychology*, 2, 113–129.

Rahman, A. (1991). *Structural Adjustment and Income Distribution*. Geneva: ILO.

Ralston, D. A., Gustafson, D. J., Elsass, P. M., Cheung, F., and Terpstra, R. H. (1992). Eastern values: a comparison of managers in the United States, Hong Kong, and the People's Republic of China. *Journal of Applied Psychology*, 77, 664–671.

Rao, G. L. (1979). *Brain Drain and Foreign Students: A Study of the Attitudes and Intentions of Foreign Students in Australia, the USA, Canada, and France*. Brisbane: University of Queensland Press.

Reeve, J. M. (1992). *Understanding Motivation and Emotion*. Orlando, FL: Holt, Rinehart, & Winston Inc.

Reser, J. P. (1991). Aboriginal mental health: conflicting cultural perspectives. In J. Reid and P. Trompf (eds), *The Health of Aboriginal Australia* (pp. 218–291). Marrickville, Australia: Harcourt Brace Jovanovich.

Richardson, B. (1987). Psychology in Papua New Guinea: a brief overview. In G. H. Blowers and A. M. Turtle (eds), *Psychology Moving East: The Status of Western Psychology in Asia and Oceania* (pp. 289–303). Sydney: Sydney University Press.

Richardson, J. T. E. (1994). Cultural specificity of approaches to study in higher education: a literature survey. *Higher Education*, 27, 449–468.

Richardson, J. T. E., Landbeck, R., and Mugler, F. (1995). Approaches to study in higher education: a comparative study in the South Pacific. *Educational Psychology*, 15, 417–432.

Ridker, R. G. (1994). *The World Bank's Role in Human Resource Development in Sub-Saharan Africa: Education, Training, and Technical Assistance*. Washington, DC: World Bank.

Rist, R. C. (1995). Postscript: development questions and evaluation answers. *New Directions for Evaluation*, 67, 167–174.

Rivera, A. N. (1984). *Hacia una psicoterapia para el Puertorriqeño*. San Juan, Puerto Rico: CEDEPP.

Robbins, S. P., Waters-Marsh, T., Cacioppe, R., and Millett, P. (1994). *Organisational Behaviour: Concepts, Controversies, and Applications*. Sydney: Prentice-Hall.

Robertson, I. T., Gratton, L., and Rout, U. (1990). The validity of situational interviews for administrative jobs. *Journal of Organizational Behaviour*, 11, 69–76.

Rodgers, R., and Hunter, J. E. (1991). Impact of management by objectives on organizational productivity. *Journal of Applied Psychology*, 76, 322–336.

Rodriques, A., and Assmar, E. M. L. (1988). On some aspects of distributive justice in Brazil. *Revista interamericana de psicología*, 22, 1–20.

Rogers, E. M. (1971). *Communication of Innovations: A Cross–Cultural Approach* (2nd edn). New York: Free Press/Collier-Macmillan.

Rondinelli, D. A. (1986). Improving development management: lessons from the evaluation of USAID projects in Africa. *International Review of Administrative Sciences*, 52, 421–445.

Rosenhan, D. L. (1973). On being sane in insane places. *Science*, 179, 250–258.

Rosenthal, R. (1968). Self-fulfilling prophecy. *Psychology Today*, 2, 44–51.

Ross, L. (1977). The intuitive psychologist and his shortcomings: distortions in the attribution process. *Advances in Experimental Social Psychology*, 10, 174–221.

Ross, L., and Nisbett, R. E. (1991). *The Person and the Situation: Perspectives of Social Psychology*. New York: McGraw-Hill.

Roth, A. E., Prasnikar, V., Okuno-Fujiwara, M., and Zamir, S. (1991). Bargaining and market behavior in Jerusalem, Ljubljana, Pittsburgh, and Tokyo: an experimental study. *American Economic Review*, 81, 1068–1095.

Rubin, J. Z., and Brown, B. (1975). *The Social Psychology of Bargaining and Negotiations*. New York: Academic Press.

Rubin, Z., and Peplau, A. (1973). Belief in a just world and reactions to another's lot: a study of participants in the national draft lottery. *Journal of Social Issues*, 29, 73–93.

Rubin, Z., and Peplau, A. (1975). Who believes in a just world? *Journal of Social Issues*, 31, 65–89.

Rugimbana, R. (1996). *A Case of Inequitable Salaries between Expatriate and Host in Tanzania*. Newcastle, Australia: University of Newcastle, Department of Management Case Studies.

Rugimbana, R., Zeffane, R., and Carr, S. C. (1996). Marketing psychology in developing countries. In S. C. Carr and J. F. Schumaker (eds), *Psychology and the Developing World* (pp. 140–149). Westport, CT: Praeger.

Ryan, W. (1971). *Blaming the Victim*. New York: Vintage.

Sahara, T. (1991). Donor delivery style, learning, and institutional development. *Journal of Management Development*, 10, 60–83.

Salmen, L. F. (1995). The listening dimension of evaluation. *New Directions for Evaluation*, 67, 147–154.

Sánchez, E. (1996). The Latin American experience in community social psychology. In S. C. Carr and J. F. Schumaker (eds), *Psychology and the Developing World* (pp. 119–129). Westport, CT: Praeger.

Sarajar, T., and Soemitro, D. (1997). *Attributions for Poverty among Jakarta's Rich and Poor*. Depok, Indonesia: Department of Social Psychology, Universitas Indonesia.

Sarson, S. B. (1974). *The Psychological Sense of Community: Prospects for a Community Psychology*. San Francisco: Jossey-Bass.

Savery, L. K., and Swain, P. A. (1985). Leadership style: differences between expatriates and locals. *Leadership and Organization Development Journal*, 6, 8–11.

SBS (Social Broadcasting Service) (1996). *The Cutting Edge: The Ghost of Health in the World Health Organisation*. July 30.

Schiffman, H. R. (1982). *Sensation and Perception*. New York: John Wiley and Sons.

Schmitt, D. R., and Marwell, G. (1972). Withdrawal and reward allocation as responses to inequity. *Journal of Experimental Social Psychology*, 8, 207–221.

Schneider, S. (1988). National vs. corporate culture: implications for Human Resource Management. *Human Resource Management*, 27, 231–246.

Schneider, S. C. (1991). National vs. corporate culture: implications for human resource management. In M. Mendenhall and G. Oddou (eds), *Readings and Cases in International Human Resource Management* (pp. 13–27). Boston, MA: PWS Kent Publishing Company.

Schoeck, H. (1969). *Envy: A Theory of Social Behaviour*. London: Secker & Warburg.

Schopper, D., Van Praag, E., and Kalibala, S. (1996). Psycho-social care for AIDS patients in developing countries. In S. C. Carr and J. F. Schumaker (eds), *Psychology and the Developing World* (pp. 173–179). Westport, CT: Praeger.

Schultz, K. (1991). Women's adult development: the importance of friendship. *Journal of Independent Social Work, 5*, 19–30.

Schwartz, S. H. (1992). Universals in the content and structure of values: theoretical advances and empirical tests in 20 countries. *Advances in Experimental Social Psychology, 25*, 1–65.

Schwartz, S. H. (1997). Values and culture. In D. Munro, J. F. Schumaker, and S. C. Carr (eds), *Motivation and Culture* (pp. 69–84). New York: Routledge.

Schwarzwald, J., Koslowsky, M., and Shalit, B. (1992). A field study of employees' attitudes and behaviors after promotion decisions. *Journal of Applied Psychology, 77*, 511–514.

Seddon, J. (1985). The development and indigenisation of Third World business: African values in the workplace. In V. Hammond (ed.), *Current Research in Management* (pp. 98–109). London: Pinter.

Selmer, J. (1996). What do expatriate managers know about their HCN subordinates' work values: Swedish executives in Hong Kong. *Journal of Transnational Management Development, 2*, 5–20.

Semler, R. (1994). *Maverick: The Success Story behind the World's most Unusual Workplace.* London: Arrow.

Senge, P. M. (1992). *The Fifth Discipline.* Sydney: Random House.

Shaba, B., MacLachlan, M., Carr, S. C., and Ager, A. (1993). Palliative versus curative beliefs regarding tropical epilepsy as a function of medical and traditional attributions. *Central African Journal of Medicine, 39* (8), 65–67.

Shamir, B. (1990). Calculations, values, and identities: the sources of collectivistic work motivation. *Human Relations, 43*, 13–32.

Shaw, J., and Clay, E. (1993). *World Food Aid: Experiences of Recipients and Donors.* New York: Heinemann.

Shaw, M. E. (1964). Communication networks. *Advances in Experimental Social Psychology, 1*, 111–147.

Sheatsley, P. B. (1983). Questionnaire construction and item writing. In P. H. Rossi, J. D. Wright, and A. B. Anderson (eds), *Handbook of Survey Research* (pp. 195–230). New York: Academic Press.

Sheridan, M. (1976). Young women leaders in China. *Signs: Journal of Women in Culture and Society, 2*, 59–89.

Sherif, M. (1936). *The Psychology of Social Norms.* New York: Harper & Row.

Sherif, M. (1937). An experimental approach to the study of attitudes. *Sociometry, 1*, 90–98.

Sherif, M. (1956). Experiments in group conflict. *Scientific American, 195*, 54–58.

Shouksmith, G. D. (1996). History of psychology in developing countries. In S. C. Carr and J. F. Schumaker (eds), *Psychology and the Developing World* (pp. 15–25).

Shumaker, S. A., and Jackson, J. S. (1979). The aversive effects of nonreciprocated benefits. *Social Psychology Quarterly, 42*, 148–158.

Shweder, R. A., and Bourne, E. J. (1982). Does the concept of the person vary cross-culturally? In A. J. Norsello, and G. M. White (eds), *Cultural Conceptions of Mental Health and Therapy* (pp. 97–137). Boston, MA: Reidel Publishing Company.

Silverman, P., and Maxwell, R. J. (1978). How do I respect thee? Let me count the

237

ways: deference towards elderly men and women. *Behavior Science Research, 13,* 91–108.

Simukonda, H. H. M. (1992). The NGO sector in Malaŵi's socio-economic development. In G. C. Z. Mhone (ed.), *Malaŵi at the Crossroads: The Post-Colonial Political Economy* (pp. 298–348). Harare: Sapes Books.

Singer, H. (1991). Foreword. In G. Standing, and V. Tokman (eds), *Towards Social Adjustment: Labour Market Issues in Structural Adjustment.* Geneva: ILO.

Singh, S., and Vasudeva, P. (1977). A factorial study of the perceived reasons for poverty. *Asian Journal of Psychology and Education, 2,* 51–56.

Sinha, D. (1989a). Cross-cultural psychology and the process of indigenisation: a second view from the Third World. In D. M. Keats, D. Munro, and I. Mann (eds), *Heterogeneity in Cross-Cultural Psychology* (pp. 24–40). Lisse, Netherlands: Swets & Zeitlinger.

Sinha, D. (1989b). Interventions for development out of poverty. In R. W. Brislin (ed.), *Applied Cross-Cultural Psychology* (pp. 77–97). Newbury Park, CA: Sage.

Sinha, D. (1990). Interventions for development out of poverty. In R. W. Brislin (ed.), *Applied Cross-Cultural Psychology* (pp. 77–97). Newbury Park, CA: Sage.

Sinha, D. (1991). Rise in the population of the elderly, familial changes, and their psychosocial implications: the scenario of the developing countries. *International Journal of Psychology, 26,* 633–647.

Sinha, D., and Holtzman, W. (eds). (1984). The impact of psychology on Third World development. *International Journal of Psychology* (whole issue).

Sinha, J. B. P. (1990). A model of effective leadership styles in India. In A. M. Jaeger, and R. N. Kanungo (eds), *Management in Developing Countries* (pp. 252–263). London: Routledge.

Sink, D. S. (1983). Using the Nominal Group technique effectively. *National Productivity Review, 2,* 181.

Skitka, L. J., McMurray, P. J., and Burroughs, T. E. (1991). Willingness to provide post-war aid to Iraq and Kuwait: an application of the contingency model of distributive justice. *Contemporary Social Psychology, 15,* 179–188.

Sloan, T. S. (1990). Psychology for the Third World? *Journal of Social Issues, 46,* 1–20.

Sloan, T. S., and Montero, M. (eds). (1990). Psychology for the Third World: a sampler. *Journal of Social Issues, 46* (whole issue), 1–165.

Smith, B., and Carr, S. C. (1997). Number of achievements and impression formation: the motivational gravity dip. *South Pacific Journal of Psychology, 9,* 7–19.

Smith, K. B., and Stone, L. H. (1989). Rags, riches and bootstraps: beliefs about the causes of wealth and poverty. *The Sociological Quarterly, 30,* 93–107.

Smith, P. B., and Bond, M. H. (1993). *Social Psychology across Cultures.* Hemel Hempstead: Harvester-Wheatsheaf.

Snyder, M., and Omoto, A. M. (1992). Volunteerism and society's response to the HIV epidemic. *Current Directions in Psychological Science, 1,* 113–116.

Solomon, R. L., and Corbit, J. D. (1974). An opponent–process theory of motivation: I. Temporal dynamics of affect. *Psychological Review, 81,* 119–145.

Southern African Economist (1992). The high cost of AIDS. *Southern African Economist, 5,* 14–17.

Squire, L. (1995). Evaluating the effectiveness of poverty alleviation programs. *New Directions for Evaluation, 67,* 27–37.

Srinivas, K. M. (1995). Organization development for national development: a review of evidence. In D. Saunders and R. N. Kanungo (eds), *Employee Management in Developing Countries* (pp. 197–223). Greenwich, CT: JAI Press.

Statt, D. A. (1994). *Psychology and the World of Work*. London: Macmillan.

Stepina, L. P., and Perrewe, P. L. (1987). The impact of inequity on task satisfaction, internal motivation, and perceptions of job characteristics. *Journal of Social Behavior and Personality*, 2, 117–125.

Stone, L. (1992). Cultural influences in community participation in health. *Social Science and Medicine*, 35, 409–417.

Storms, M. D. (1973). Videotape and the attribution process: Reversing actors' and observers' point of view. *Journal of Personality and Social Psychology*, 27, 165–175.

Stouffer, S. A., Suchman, E. A., De Vinney, L. L., Star, S. A., and Williams, R. M. (1949). *The American Soldier: Adjustment During Army Life, Vol 11*. Princeton, NJ: Princeton University Press.

Summers, J. P., and Hendrix, W. H. (1991). Modelling the role of pay equity perceptions: a field study. *Journal of Occupational Psychology*, 64, 145–157.

Taffinder, P. A., and Viedge, C. (1987). The Nominal Group Technique in management training. *Industrial and Commercial Training, July/August*, 16–20.

Tajfel, H. (1978). *Differentiation between Social Groups: Studies in Intergroup Behaviour*. London: Academic Press.

Tamayo, A. (1994). Factorial scale for causal attributions of poverty. *Psicología teoria e pesquisa*, 10, 21–29.

Tan, E. S., and Lipton, G. (1993). Recent developments and trends in mental health care in the Western Pacific region. *International Journal of Mental Health*, 22, 22–35.

Tattam, A. (1996). Indigenous nurses. *Australian Nursing Journal*, 3, 18–19.

Tauber, R. T. (1997). *Self-Fulfilling Prophecy: A Practical Guide to its Use in Education*. Westport, CT: Praeger.

Taylor, F. W. (1912). *Scientific Management*. New York: Harper and Row.

Taylor, M., and Veale, A. (1996). Rethinking the problem of streetchildren: parallel causes and interventions. In S. C. Carr and J. F. Schumaker (eds), *Psychology and the Developing World* (pp. 90–99). Westport, CT: Praeger.

Taylor, R., and Yavalanavanua, S. (1997). Linguistic relativity in Fiji: a preliminary study. *South Pacific Journal of Psychology*, 9.

Taylor, S. E., and Fiske, S. T. (1975). Point of view and percpetions of causality. *Journal of Personality and Social Psychology*, 32, 439–435.

Templer, A., Beaty, D., and Hofmeyr, K. (1992). The challenge of management development in S. Africa: so little time and so much to do. *Journal of Management Development*, 11, 32–41.

Thibaut, J. W., and Kelley, H. H. (1959). *The Social Psychology of Groups*. New York: John Wiley and Sons.

Thomas, A. (1996). What is development management? *Journal of International Development*, 8, 95–110.

Thomas, D. R. (1994). Developing community and social psychology for Aotearoa: experiences from a New Zealand programme of indigenization. Third Afro-Asian Congress of Psychology, Bangi, Selangor, Malaysia, August.

Thomas, D. R., and Nikora, L. W. (1996). Maori, Pakeha, and New Zealander: ethnic and national identity among New Zealand students. *Journal of Intercultural Studies*, *17*, 29–39.

Thompson, A. A., and Strickland, A. J. (1990). *Strategic Management: Concepts and Cases*. Boston, MA: BPI Irwin.

Thornton, B., Kirchner, G., and Jacobs, J. (1991). Influence of a photograph on a charitable appeal: a picture may be worth a thousand words when it has to speak for itself. *Journal of Applied Social Psychology*, *21*, 433–445.

Todaro, M. (1994). *Economic Development*. London: Longman.

Tornow, W. W. (1993). Perception or reality: is multi-perspective measurement a means or an end? *Human Resource Management*, *32*, 221–230.

Townsend, P. (1979). *Poverty in the United Kingdom: A Survey of Household Resources and Standards of Living*. Harmondsworth: Penguin.

Traynor, W. J., and Watts, W. R. (1992). Management development in the Pacific during the 1990s: how to survive with coconuts. *Journal of Management Development*, *11*, 67–79.

Trebilcock, A. (1996) Structural adjustment and tripartite consultation. *The International Journal of Public Sector Management*, *9* (1).

Triandis, H. C. (1978). Some universals of social behavior. *Personality and Social Psychology Bulletin*, *4*, 1–16.

Triandis, H. C. (1980). Value, attitudes, and intergroup behavior. In M. M. Page (ed.), *Nebraska Symposium on Motivation, Beliefs, Attitudes, and Values, Vol. 1* (pp. 195–259). Lincoln, NE: University of Nebraska.

Triandis, H. C., McCusker, C., and Hui, C. H. (1990). Multimethod probes of individualism and collectivism. *Journal of Personality and Social Psychology*, *59*, 1006–1020.

Trompenaars, F. (1993). *Riding the Waves of Culture: Understanding Cultural Diversity in Business*. London: Brearley.

Turner, J. C. (1975). Social comparison and social identity: some prospects for intergroup behavior. *European Journal of Social Psychology*, *5*, 5–34.

Turner, J. C. (1991). *Social Influence*. Milton Keynes: Open University Press.

Tversky, A., and Kahneman, D. (1973). Availability: a heuristic for judging frequency and probability. *Cognitive Psychology*, *5*, 207–232.

Tversky, A., and Kahneman, D. (1981). The framing of decisions and the psychology of choice. *Science*, *211*, 453–458.

UNDP and World Bank (1985). *Somalia: Report of a Joint Technical Cooperation Assessment Mission* (pp. 17–18).

UNICEF, WHO, UNESCO, and UNFPA. (1993). *Facts for Life: A Communication Challenge*. Oxford: P & LA.

United Nations (1993). *Situation Analysis of Poverty in Malawi*. Lilongwe, Malawi: United Nations.

Ustun, T. B., and Sartorius, N. (eds). (1995). *Mental Illness in General Health Care: An International Study*. Geneva: World Health Organization.

Valenzi, E. R., and Andrews, I. R. (1971). Effect of hourly overpay inequity when tested with a new induction procedure. *Journal of Applied Psychology*, *55*, 22–27.

Vallance, S. (1996). The human resource crisis in Fiji's public sector: trouble in paradise. *Public Administration and Development*, *16*, 91–103.

Van de Walle, N. (1995). The politics of effectiveness. In S. Ellis (ed.), *Africa Now: People, Policies, Institutions*. London: James Currey.

Vaughan, G. M. (1978). Social categorization and intergroup behaviour in children. In H. Tajfel (ed.), *Differentiation between Social Groups* (pp. 339–360). London: Academic Press.

Verhelst, T. G. ([1987] 1990). *No Life Without Roots: Culture and Development.* (trans. Bob Cumming). London: Zed Books.

Vittitow, D. (1983). Applying behavioural science to Third World development. *Journal of Applied Behavioural Science, 19,* 307–317.

Wagner, D. A. (1996). using social science for literacy work in developing countries. In S. C. Carr and J. F. Schumaker (eds), *Psychology and the Developing World* (pp. 81–89). Westport, CT: Praeger.

Wang, T., Brownstein, R., and Katzev, R. (1989). Promoting charitable behaviour with compliance techniques. *Applied Psychology: An International Review, 38,* 165–183.

Wapenhans, W. (1993). *Effective Implementation: Key to Development Impact.* Washington, DC: World Bank.

Ward, C., and Kennedy, A. (1992). Locus of control, mood distrubance, and social difficulty during cross-cultural transitions. *International Journal of Intercultural Relations, 16,* 175–194.

Warr, P. (1982). A national study of non-financial employment commitment. *Journal of Occupational Psychology, 55,* 297–312.

Watkins, D., Akande, A., and Mpofu, E. (1994). The assessment of learning processes: some African data. *Ife Psychologia: An International Journal, 2,* 1–18.

Watters, P. A., Ball, P. J., and Carr, S. C. (1996). Social processes as dynamical processes: qualitative dynamical systems theory in social psychology. *Current Research in Social Psychology, 1*(7), 60–68.

Weick, K. E. (1984). Small wins: redefining the scale of social problems. *American Psychologist, 39,* 40–49.

Weiner, B. (1980). The role of affect in rational (attributional) approaches to human motivation. *Educational Research, July–August,* 4–11.

Weiner, B. (1991). Metaphors in motivation and attribution. *American Psychologist, 46,* 921–930.

Weisman, C. S., Gordon, D. L., Cassard, S. D., and Bergner, M. (1993). The effects of unit self-management on hospital nurses' work process, work satisfaction, and retention. *Medical Care, 31,* 381–393.

Whaples, G. C., and Ogunfiditimi, T. O. (1979). Construction of a five-factor attitude instrument for evaluation of the food for work programs. *Journal of Social Psychology, 107,* 291–292.

Wheeler, L., Deci, E. L., Reis, H. T., and Zuckerman, M. (1978). *Interpersonal Influence.* Boston, MA: Allyn & Bacon.

WHO Report (1996). Poverty and ill-health in developing countries: learning from NGOs. World Health Organization and Government of Ireland, 12–14 June.

Wilke, H. and Van Knippenberg, A. (1990). Group performance. In M. Hewstone, W. Stroebe, J. P. Codol, and G. M. Stephenson (eds), *Introduction to Social Psychology* (pp. 315–349). Oxford: Basil Blackwell.

Williams, G. (1993). All for Health: A Resource Book for Facts for Life. New York: UNICEF.

Wilson, A. (1987). Approaches to learning among Third World tertiary science students: Papua New Guinea. *Research in Science and Technological Education, 5,* 59–67.

Wilson, D., Zenda, A., McMaster, J., and Lavelle, S. (1992). Factors predicting Zimbabwean students' intentions to use condoms. *Psychology and Health, 7,* 99–114.

Wong, M. M. L. (1996). Shadow management in Japanese companies in Hong Kong. *Asia Pacific Journal of Human Resources, 34,* 95–110.

World Bank (1994). *The World Bank and Participation.* Washington, DC: The World Bank.

World Bank. (1995a). *Strengthening the Effectiveness of Aid: Lessons for Donors.* Washington, DC: World Bank.

World Bank (1995b). *World Bank Report.* Washington, DC: World Bank.

World Bank News (1996). Success and failure in Africa – two case studies: technical assistance in Ghana and Uganda. *Development Bulletin, 37,* 56–57.

World Bank Participation Sourcebook (1995). *Environmentally Sustainable Development.* Washington, D.C: World Bank.

Zeira, Y., and Banai, M. (1984). Present and desired methods of selecting expatriate managers for international assignments. *Human Resources Management, 13,* 29–35.

Zheng, X., and Berry, J. W. (1991). Psychological adaptation of Chinese sojourners in Canada. *International Journal of Psychology, 26,* 451–470.

Zhurvalev, A. L., and Shorokhova, E. V. (1984). Social psychological problems of managing the collective. In L. H. Strickland (ed.), *Directions in Soviet Social Psychology.* New York: Springer.

Zimba, C. G., and Buggie, S. E. (1993). An experimental study of the placebo effect in traditional African medicine. *Behavioral Medicine, 19* (3), 103–109.

Zimbardo, P. G. (1982). Pathology of imprisonment. In D. Krebs (ed.), *Readings in Social Psychology* (pp. 249–251). New York: Harper & Row.

Zindi, F. (1996). Educational selection in developing countries. In S. C. Carr and J. F. Schumaker (eds), *Psychology and the Developing World* (pp. 71–80). Westport, CT: Praeger.

Zucker, G., and Weiner, B. (1993). Conservatism and perceptions of poverty: an attributional analysis. *Journal of Applied Social Psychology, 234,* 925–943.

Zuckerman, M. (1975). Belief in a Just World and altruistic behavior. *Journal of Personality and Social Psychology, 31,* 972–976.

Zuckerman, M., and Reis, H. T. (1978). Comparison of three models for predicting altruistic behavior. *Journal of Personality and Social Psychology, 36,* 498–510.

INDEX

Aamodt, M.G. 131
achievement 4–5
Achievement-Ascription of status
   (Trompenaars) 76, 77, 159
Achievement-in-Context 127–8
actor-observer difference in attribution
   25, 31, 35; international 33–4,
   35–6
Adams, J.S. 107, 108, 110, 111, 150,
   151
Adamson, P. 138, 170–3 *passim*
affect, social *see* social affect
Africa: aid chain and work in 46–7,
   53–4, 64; educational aid 142;
   expatriate work motivation 102, 108;
   and global market 91; intercultural
   work dynamics 163, 164, 202; 'Pay
   Me!' reaction 188, 189; traditional/
   biomedical health services 178; *see also*
   individual countries
African Americans: attributions of
   poverty 35; work motivation,
   dynamics 111, 157
Ager, A.K. 94–6, 176, 179
agricultural research, lack of
   coordination 51–2
aid agendas, establishing 46–50
aid chains 14–15, 44–64, 192, 200;
   communication process 51–5;
   establishing agenda 46–50;
   groupthink 15, 50–1, 64;
   participation 15, 55–64
aid cycle 11–12, 206; applying
   psychology to 13–19
aid donor/host relationship *see* donor/
   host relationship
aid for trade 37, 158, 199
aid projects: entropy 19, 204; failure/

success xii, 2, 3, 19, 102, 136; *see also*
   effectiveness of aid
aid relief 37, 73, 94, 187
aid scholarships 17–18, 121–34, 136,
   137, 141, 142, 201
aid training 81, 139; *see also* educational
   aid *and under* expatriates, donor; hosts
   at home
aid-tying 47, 59
AIDS: cognitive tolerance in treatment
   and prevention 172–3, 182–3;
   community research 193
AIDS prevention 7, 30, 56, 62, 131–2;
   transactional positioning 137–8, 139,
   140, 143
Airhihenbuwa, C. 175
Ajzen, I. 40
Akhan, Akhter Hameed 84–5
Alatas, S.H. 156
'All for Health' (Williams/UNICEF)
   170–1
all-round 360-degree feedback 68, 79,
   80, 143
Allen, T. xi
altruism: donor 40; expatriate work
   motivation 104, 106, 108, 113, 201;
   and reciprocity 196
Alvarez, H. 63
Anderson, N.R. 124
approach-avoidance tendencies 158
Approaches to Study Inventory (ASI)
   127
appropriate technology 84
archetypes, system 11–12, 202–3
Aronson, E. 164
Asia 102, 163, 167, 189; values 5, 77;
   see also East Asia
assessment stage 67–8